Time and the Tuolumne Landscape

TIME AND THE TUOLUMNE LANDSCAPE

Continuity and Change in the Yosemite High Country

Thomas R. Vale

Geraldine R. Vale

UNIVERSITY OF UTAH PRESS
Salt Lake City, Utah

Printed on acid-free paper

Library of Congress Cataloging-in-Publication Data

Vale, Thomas R., 1943–
Time and the Tuolumne landscape : continuity and change in the Yosemite high country / Thomas R. Vale, Geraldine R. Vale.
p. cm.
Includes bibliographical references (p.) and index.
ISBN 0-87480-429-9 (acid-free paper)
1. Landforms—California—Tuolumne River Watershed. 2. Landforms—California—Yosemite National Park. 3. Climatic changes—California—Tuolumne River Watershed. 4. Climatic changes—California—Yosemite National Park. 5. Yosemite National Park (Calif.) I. Vale, Geraldine R., II. Title.
GB428.C3V35 1994
508.794'47—dc20 93-21700

For our parents,

Una and Jimmy Vale

Catharine and Joseph Depta,

who first introduced us

to the mountains

CONTENTS

PREFACE

Watery sunlight wends its way through the summer haze hanging over this Midwestern college town and filters through the double panes of my study's window. It comes to rest on a small photograph, a print from an early slide, that I hold in my hands. Gently, I touch the figures in the image.

The setting of the picture is 2,000 miles westward, the elevation 8,000 feet skyward, on the eastern side of Yosemite National Park. Clustered along the bank of the Tuolumne River at a spot but a short walk from the Tuolumne Meadows Campground, are a mother and her three children. Their attention seems drawn across the downcut channel and tree-bordered meadow upward through the thin, clear air, toward Lembert Dome and the cloud puffs suspended above it. The date is July 1952. The young family is that of my childhood; the landscape is one dear not only to me and to my family but to generations of Americans.

Even though for nearly two-score years the fragile grasses have been well trampled and the crystal waters heavily angled, the Tuolumne landscape in the photo should be readily identifiable by any of the faithful who yearly flock to their beloved Yosemite high country. The stability of Lembert Dome—the great rock's massive form, the details of fractures and erratic boulders, the dark patches of lichen on the white granite, the few trees scattered across its slope and along its summit—ensures this easy recognition.

Other natural features in the photograph also stir nostalgia in the Tuolumne enthusiast—the framing forms of lodgepole pine, the foreground sweep of short, green meadow grasses, the gently rolling waters of the Tuolumne River, even the cumulus clouds in the blue sky overhead. Less constant than the indomitable face of Lembert Dome, these features vary more through time, whether by centuries, seasons, or seconds. The trees grow taller and more numerous today than in 1952; the expanse of meadow in the 1980s correspondingly retreats under pressure of the encroaching forest. The river in the photograph flows gently during the vanishing snowmelt days of midsummer, in stark contrast to its roiling a month or two earlier, floodwater-fed from the deep snowpacks of winter. Especially transient, the pattern of clouds kaleidoscopes slowly but ceaselessly across the sky, interrupted only by the click of a shutter.

Between the time of the photograph and today, then, the landscape of Tuolumne evinces reassuring persistence as well as intriguing change. The same might be said of my family's journeys to this special place in the middle of the century as compared to those of a typical family today. In 1952, we approached Tuolumne over twenty-one slow miles of the steep and twisting old Tioga Road, now mostly abandoned or rebuilt, yet we also sped along on the newer segments of road west of White Wolf and east of Cathedral Creek, as do today's visitors. We easily secured a favorite tent site in the campground touted by the Park Service in 1952 as so large that space was always available; today vacationers typically flood Ticketron with reservations months in advance, guarding against the "Campground Full" sign raised at the monitoring kiosk almost each summer evening. We strolled through the owl's clover and knotweed of Tuolumne Meadows, lured by vivid sunsets, grazing deer, and soft evening breezes, much as do families today, though the trails then were less worn by foot and hoof, the silence less broken by the cry of voices or the hum of motors.

Intrigued by such personal reflections, my wife and I decided to explore further the theme of persistence and change in the Tuolumne landscape, but as it has developed over eighty, rather than forty, years. We pursued a "repeat photography" study, based on a set of about 80 views originally photographed around the turn of the century and rephotographed by us from 1984 to 1988. (An additional three dozen pairs were taken but are not included in this book.) The original photos are part of early geological research in Yosemite, and the majority of the scenes were taken by G. K. Gilbert (the "father of modern landform science"), who started intermittent studies in Yosemite in 1903, and F. E. Matthes, whose 1930 *Geologic History of the Yosemite Valley* was the culmination of years of research in the Park. These photographs are part of the collection of the U.S. Geological Survey Photographic Library in Denver, Colorado. A few additional old photos (identified as from the National Park Service) are from the Research Library in Yosemite Valley. The photo pairs in the following pages are grouped into chapters by the predominating subject, such as vegetation, but a table identifies overlapping subjects (Table 1). A map allows the reader to locate specific geographic places (Figure 1). Our topical divisions stress the landscape, the physical appearance of the Earth, but we also use the photos as stimuli to think about changes in human experiences over the eighty years bracketed by the photo pairs. Our book, then, is an exploration of continuity and change in the land and life of Tuolumne, and we hope that it encourages lovers of the Yosemite high country to contemplate the importance of time in this special landscape.

T. Vale
Madison, Wisconsin

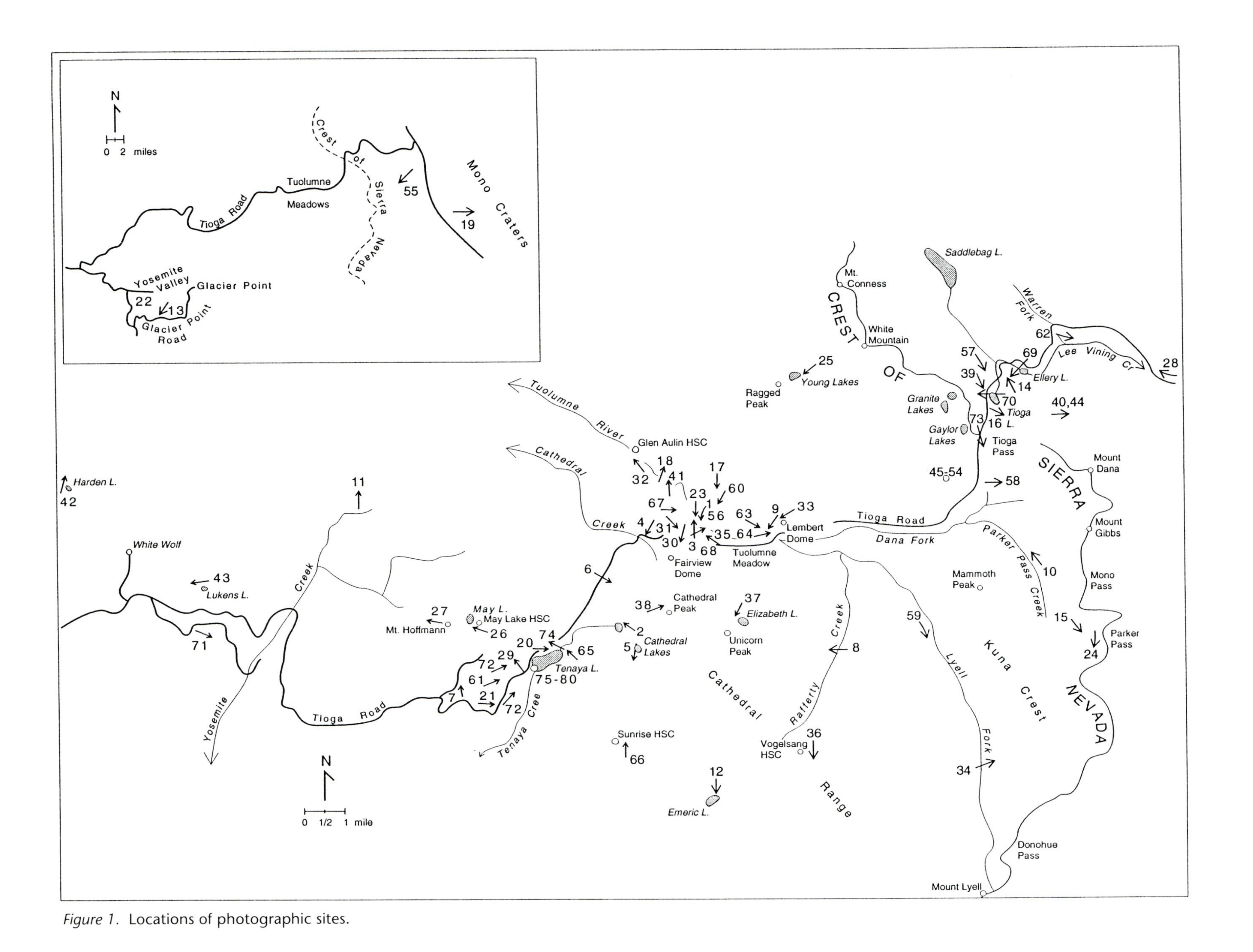

Figure 1. Locations of photographic sites.

ACKNOWLEDGMENTS

One summer of fieldwork for this project was supported by a grant from the American Philosophical Society. Additional support was provided by the National Park Service, through the efforts of Dr. Jan Van Wagtendonk, Research Scientist at Yosemite National Park. We appreciate the help of personnel at the Research Library of the National Park Service in Yosemite and at the U.S. Geological Survey Photographic Library in Denver, Colorado.

ONE

Change in the Landscape

In the summer of 1868, John Muir first walked across the hot, golden stretches of the low-lying Central Valley of California and climbed into the cooler grey and green reaches of the Yosemite Sierra. He spent ten days in Yosemite Valley that year, to return in 1869 for his famed "first summer in the Sierra." That these initial journeys propelled Muir into conservation immortality is fairly common knowledge; that they also led him into scientific controversy is less well appreciated.

At the time, the most widespread explanation for the origin of Yosemite Valley was that of the eminent geologist J. D. Whitney, who argued that the valley's vertical walls were the product of great faulting, which had split the earth apart and sent the valley floor crashing downward in a "wreck of matter and the crush of worlds" (Colby 1960). Muir, on the other hand, even in his first summer in the Yosemite landscape, recognized that slowly grinding glaciers, rather than tumultuous earthquakes, were responsible for much of the molding of the Sierra Nevada and even of the great Valley: "[T]he formation of Yosemite, . . . together with all of its various domes and sculptured walls, [was] produced and fashioned by the united labors of the grand combination of glaciers which flowed over and through it." (Colby 1960). Muir's work as a careful observer, as a scientist, would demonstrate that Whitney was wrong.

The views of Whitney and Muir on the genesis of the Yosemite landscape differ not only in the primacy of agents, whether faults or glaciers, but also in the vision of the pace and duration of change involved with the agents, whether catastrophic and sudden or gradual and continuous. Moreover, Muir expanded his view to see change in the natural world generally as ever present and ongoing:

> Nature is ever at work building and pulling down, creating and destroying, keeping everything whirling and flowing, allowing no rest but in rhythmical motion, chasing everything in endless song out of one beautiful form into another. (Muir 1901)

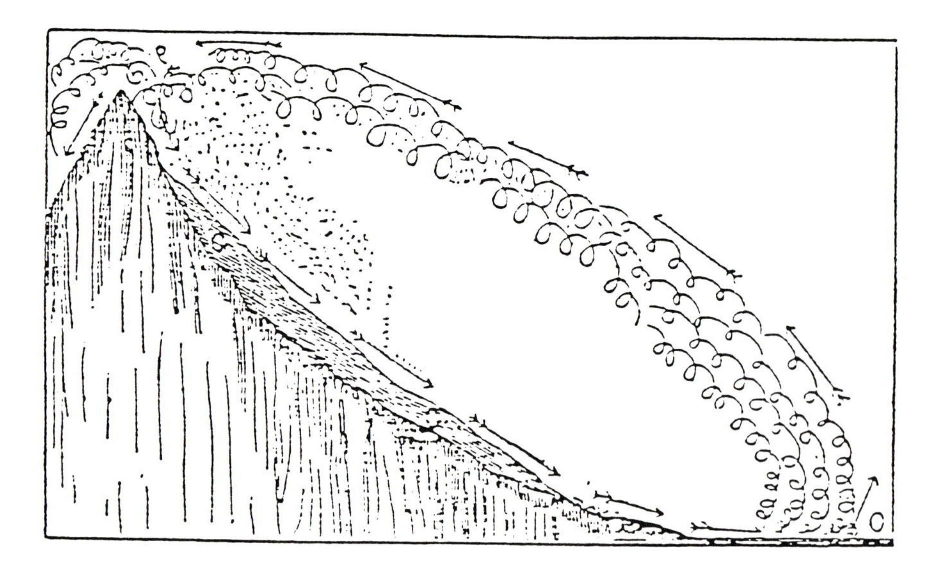

John Muir's representation of the water cycle. (SOURCE: Colby 1960)

So it also seems to those of us who frequent the wonders of Yosemite that nothing in nature seems static. Fires course across some area of Yosemite every year, as above Crane Flat in 1988, along the park's western boundary in 1990 and 1992, and on the Illilouette Creek drainage in many recent years. In Yosemite Valley, great slabs of rock sometimes crash from the cliffs, as they did beside Yosemite Falls in 1980, a November event that killed three people. Such landslides occasionally dam streams to form lakes, as one did centuries ago to create Mirror Lake. Once established, these lakes may fill with fine rock material washed into their quiet water by turbulent streams; the obvious example is Mirror Lake, now more a marsh than the lake so admired by Yosemite visitors of the last century. Influencing the intensity and frequency of both fire and erosion is the weather, the unpredictable and fluctuating precipitation and temperature, as suggested by the dry years of 1976 and 1977, followed by the record snowfall of 1983, returning to an unprecedented drought in the late 1980s and early 1990s, and then the heavy snows of 1992–93.

A vision of such a constantly changing nature is similar to an intricate timepiece, with myriad wheels spinning at varying speeds, some stopping or starting without warning, others reversing their direction, a few steady on their course. In nature, however, the "watchmaker" seems to have no preconceived notion as to how fast or how regularly the clock—or its components—should run. Individual natural phenomena behave differently. (Whitney's vision of catastrophic change may not be appropriate for the formation of Yosemite Valley, but it may be for the flanks of Mount St. Helens or the rockfall at Yosemite Falls.) A continuity of conditions or characteristics prompts the description of "static" or "stable"; a deviation invites the word "change." Having identified "change," a person might see it as occurring steadily and continuously, or only periodically. It may happen slowly or suddenly. Change might be continuing today or may have ended in the past. It might follow a cyclical pattern, with today's changes being only temporary, followed by still other changes that lead back to an initial condition. Still other change may be better described as "directional," the creation of a situation unknown in the past. Even while everything is "whirling and flowing," the eddies and currents follow countless channels. Which changes we see, which changes we comprehend, are limited by the time scale of the human lifetime, and we need to exert extra effort to free our minds of that constraint if we are to see, with Muir, that landscape change is indeed an "endless song" (Plates 1–3).

Plate 1. Fairview Dome from the western end of Tuolumne Meadows, with the summit of Pothole Dome visible above a fringe of forest. R. B. Dole 1913; Vale 1989.

As the glacial ice retreated at the end of the Ice Age, the broad lowland that we know today as Tuolumne Meadows emerged from its cold tomb into the warming sunlight. For a short time after its initial exposure, the lowland was probably a bare expanse of loose boulders, gravel, and sand. It was likely crossed by multiple streams of the meltwater-swollen Tuolumne River, each channel shifting back and forth across the plain. Forests became established quickly. In the cooler and moister times of the last several thousand years, however, rising groundwater killed the trees and allowed the establishment of the meadow grasses and flowers admired so fondly by R. B. Dole and John Muir. Over the last eighty years, lodgepole pine have invaded Tuolumne Meadows, as they have most subalpine meadows of the Sierra. In contrast to this history of recurrent change, Fairview Dome and Pothole Dome probably appear today much as they did when the melting ice first exhumed their surfaces; only weathering of some glacial polish has altered their masses over the last ten thousand years.

Plate 2. Lower Cathedral Lake from the southeast. G. K. Gilbert 1903; Vale 1984.

Lower Cathedral Lake occupies a basin scoured by glacial ice, with the water impounded behind a damming lip of hard bedrock at the far shore. The size of the lake is likely little changed since the disappearance of the glaciers, as suggested by the absence of sediment deposits on the shoreline (that is, the shore is mostly long-ago glacially scoured granitic bedrock, rather than recent sediment). On the slopes and rock exposures around the lake, the stands of lodgepole pine and mountain hemlock have thickened over the past eighty years. Invasive trees are absent from the ponds and wet meadow extending across the foreground, an area of poor drainage where an underlying bedrock shelf impedes downward movement of water. This elongated wetland was probably scoured by glacial ice, as was Lower Cathedral Lake itself, and then subsequently filled by stream gravels and sand or by growth of meadow plants.

Plate 3. Pothole Dome at the western end of Tuolumne Meadows. F. E. Matthes 1917; Vale 1987.

Glacial ice moved from right to left, riding up the resistant spine of Pothole Dome. Lodgepole pine have become established in the few fractures on the dome, where a little sand and gravel have permitted roots to hold fast. During this century, trees have increased in the meadow, although Park Service crews have removed them from the grassy lowland, except along the base of the rock. A favorite short scramble for visitors, Pothole Dome allows a leisurely walk on a glacier-scoured surface, virtually unchanged since the end of the Ice Age, as well as a striking view of life tenaciously colonizing its gentle slope. The naturalist-led group in the recent photograph is heading toward some glacial gouges on the lower flank of the dome behind the trees at the far right (see Plate 30).

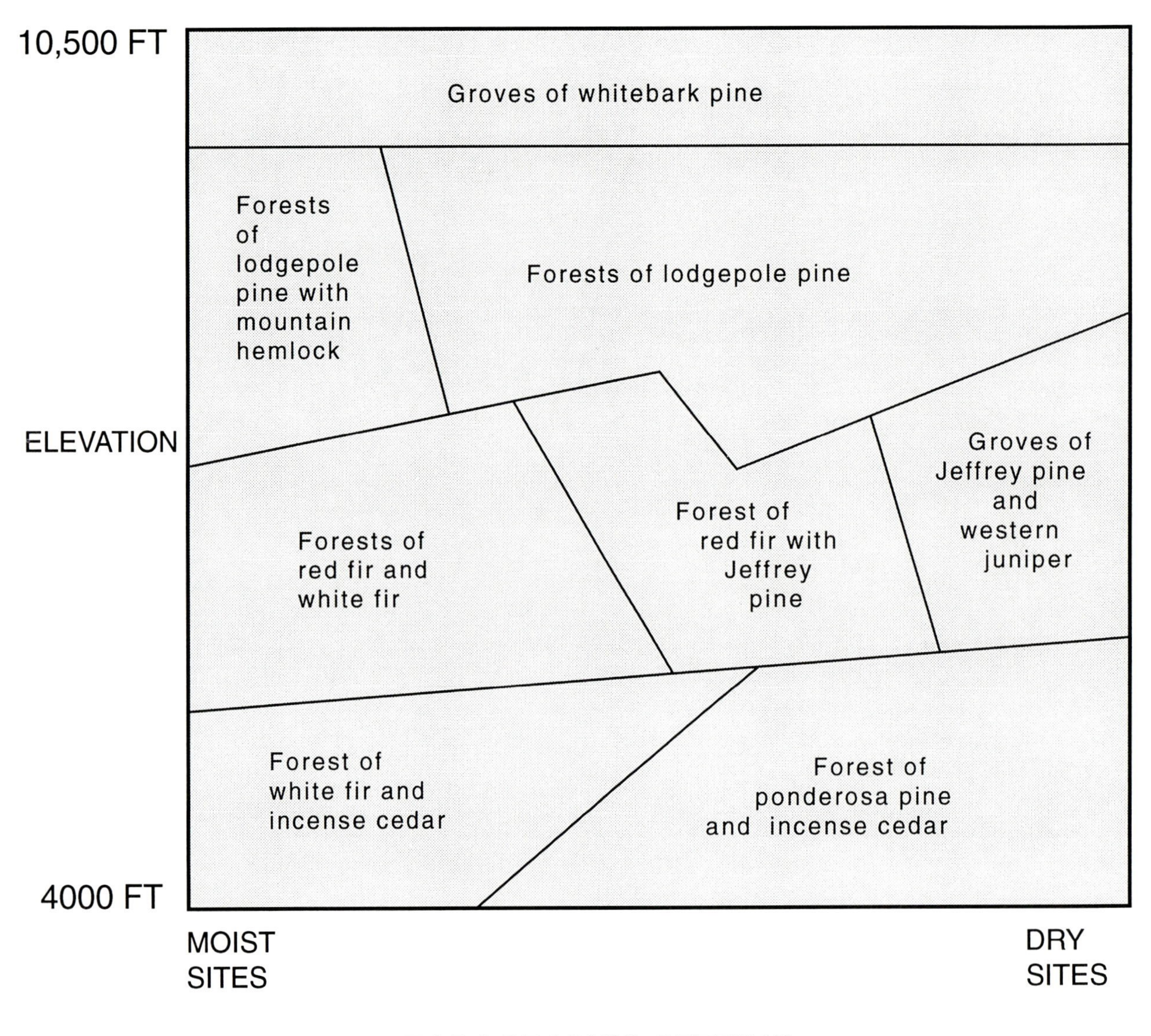

Figure 2. The generalized pattern of forest composition in Yosemite is largely determined by elevation and topographic setting, which influence not only moisture availability but also the length of the growing season. Species may also be segregated by other factors; for example, ponderosa pine and incense cedar seem to grow on more fertile soils than white fir. Disturbance events such as fire and wind throw, moreover, are influenced by characteristics of both the tree species and the nature of the physical environment. The strong links between forest composition and environmental characteristics suggest that the forest pattern is in balance or equilibrium with major environmental factors. According to this perspective, change in the vegetation results from changes in the environmental characteristics which control it. (Parker 1989.) Still, the contemporary links between these tree species and characteristics of climate and topography, as suggested by the figure, are not the only relationships possible: In late Ice Age times, forests at middle elevations in Yosemite included mountain hemlock, from modern subalpine forests, and ponderosa pine and incense cedar, from today's montane forest (Smith and Anderson 1992). Was this different forest composition reflective of the vegetation adjusting to climate change at the end of the Ice Age, or might it have been a different equilibrium forest? (SOURCE: Parker 1989)

TWO

Vegetation

The vegetation of the Tuolumne landscape creates a relatively simple pattern of forests of lodgepole pine interspersed both with meadows, sometimes wet, sometimes dry, of grasses, sedges, and flowering plants and with domes or slopes of mostly bare, smooth granite. Red fir mingles with the lodgepole in forests at slightly lower elevations; singular Jeffrey pine and Sierra juniper grace the dry, rock slopes across a wide elevation band; groves of mountain hemlock hug moist, shady recesses on the higher slopes. At the upper forest limit this motif transforms to small stands of whitebark pine, often congregating in prostrate clumps, strewn across open alpine slopes of broken rock and meadow patches.

The evergreen and needleleaf character of the trees presumably reflects advantages that such structures have in this moist, high-elevation environment (Walter 1979). The evergreen leaves can maximize photosynthesis during the short growing season because precious time is not wasted putting on new leaves, as deciduous trees would require. In addition, the presence of leaves all year allows the evergreen trees to photosynthesize during mild weather in times other than summer. The mechanically strong needleleaf form, moreover, allows the evergreen leaves to survive the long, cold winter season. Whereas most of the dominant trees in the Yosemite landscape are evergreen and needleleaf, the distributions of particular species reflect differing abilities to utilize varying amounts of moisture and energy, as suggested by the species' responses to elevation and topographic situation (Figure 2).

Trees do not dominate all sites. At the highest elevations, trees cannot form forests because of extremes in many environmental factors, including summer sun, winter cold, shallow soils, and high winds (Arno 1984). Even within the forests, particularly moist substrates, typically in valley bottoms, may be waterlogged too much of the year for the oxygen needs of trees, and nonwoody plants may then create meadow sods. Cold air that settles into valley bottoms during spring may also shorten the growing season and thus discourage tree growth. Other grassy areas, however,

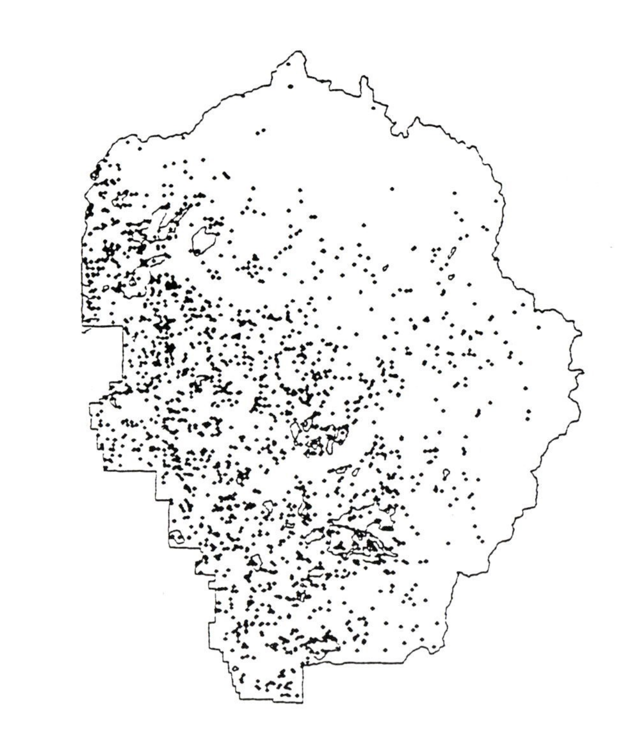

Figure 3. Lightning fires in Yosemite National Park, 1930–1983. Note that ignitions are relatively few in the higher elevations of the eastern half of the park and are common in the middle and lower elevations of the western half. Since 1972, when natural fires were first allowed to burn in Yosemite, more than four hundred lightning fires have been managed in the park. In addition, Park Service crews have ignited more than eighty other burns, designed to return forest conditions to those of pre-European times, when fires in middle and low elevations were frequent (information from National Park Service, 1990). (SOURCE: Van Wagtendonk 1986)

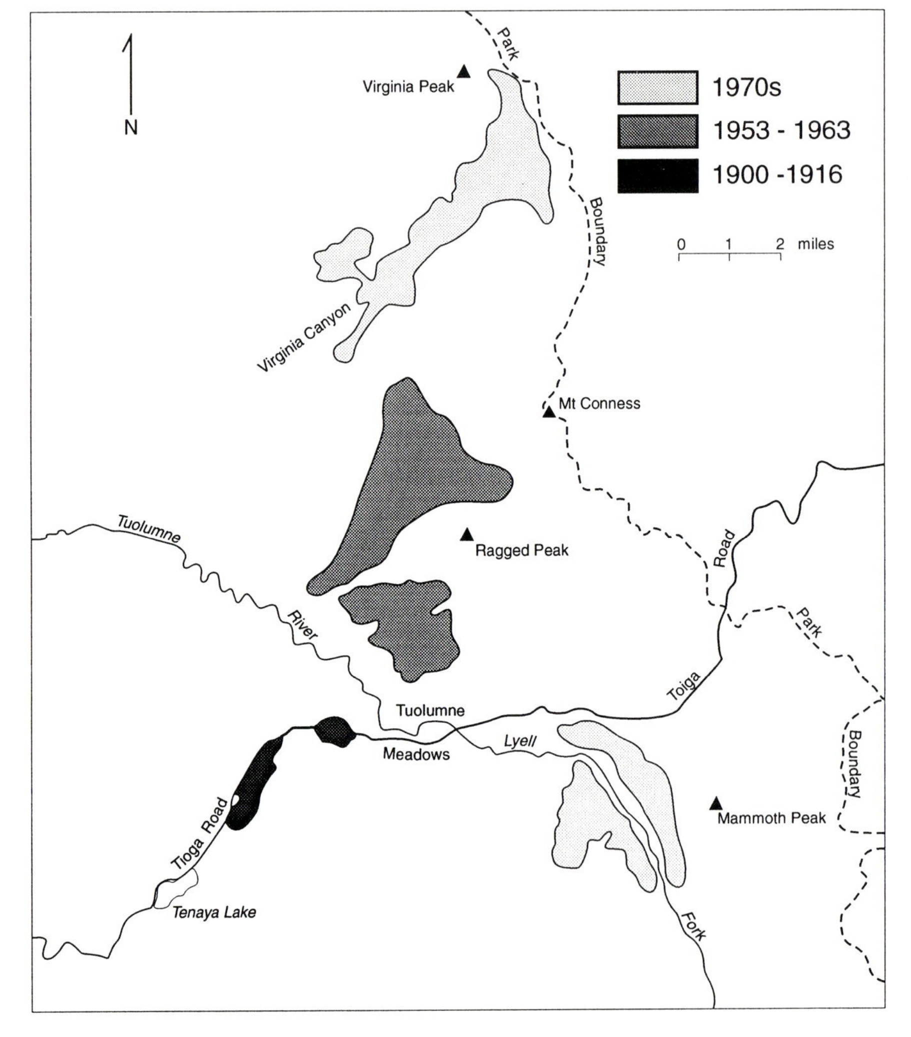

Figure 4. Periods of needle-miner moth infestations. (SOURCE: Koerber 1973)

such as those on slopes, may have particularly shallow and sandy soils, which retain too little moisture to support trees. The almost constantly dry granite domes and slickrock slopes represent an extreme of this droughty situation, and trees may germinate and persist only along cracks in the rock where moisture may soak downward and be available for plants. Overall, then, the vegetation pattern can be explained by reference to regional climate, topographic situation, and soil characteristics, and change in the pattern, according to this view, requires alteration of the environmental factors.

The pattern might be stable, but the vegetation is not static. John Muir appreciated the ubiquity of change throughout this vegetation mosaic, a transmutation which effects not discord but rather "ineffable beauty and harmony":

> The snow bends and trims the forests every winter, the lightning strikes a single tree here and there, while avalanches mow down thousands at a swoop as a gardener trims out a bed of flowers. . . . The winds go to every tree, fingering every leaf and branch and furrowed bole . . . the winds blessing the forests, the forests the winds, with ineffable beauty and harmony as the sure result. (Muir 1894/1961)

For Muir, then, the vegetation changes on short-term cycles, but these are only fluctuations in a longer-term stability.

The photo pairs support such a view of vegetation in the Tuolumne landscape, although other interpretations of change are also possible. Clearly, the fundamental vegetation pattern over the past eighty years persists, in spite of snow bending the forest trees by varying degrees each year and lightning and avalanching downing trees in certain areas in certain years. The density and species composition of the lodgepole forests and their pattern of occurrence in the landscape reveal remarkable stability (Plate 4). Even the forms of individual trees may be so little altered as to appear nearly identical in two views separated by eight decades (Plate 5).

This persistence in the vegetation reflects a lack of catastrophic change in the natural ecology of the Tuolumne high country. Great forest fires, like those of Yellowstone in 1988, do not sweep the lodgepole forests of Yosemite. The middle- and lower-elevation forests of the Sierra carry fires easily, as indicated by the 1988 burn above Crane Flat and the 1990s fires along the park's western border, but fires that are ignited by lightning in the high elevation lodgepole forests characteristically burn only small areas before dying, probably for lack of fuel (Figure 3). The single agent of sudden, widespread, and dramatic natural vegetation change might be the outbreaks of the needle-miner moth, whose larvae chew their way through the length of the needles of lodgepole pine, often contributing to the death of the trees (Figure 4; Plate 6).

Change in the vegetation in the Tuolumne landscape has nonetheless occurred during this century, and most of these changes involve an increase in the numbers or densities of trees (Vale 1987). In many locales, individual young trees have persisted and grown larger as they have matured, and this growth has been accomplished without altering the general appearance of the landscape (Plate 7). But elsewhere increases in numbers of trees have led to dramatic changes in the look of the vegetation. Most common and most conspicuous is the invasion of meadows by lodgepole pine (Plates 8–11). Even the drier, upland edges of meadows have become more forested during this century (Plate 12).

Long recognized by those interested in the high-elevation landscapes of the Sierra, these invasions of subalpine meadows by trees are usually explained by one of two hypotheses. First, climatic drying may have allowed tree establishment in locales formerly too wet for trees. Second, before park establishment, sheep grazing may have broken the cover of the meadow plants; after the trampling sheep were removed, trees could successfully germinate and thrive in this opened sod. Fire protection cannot be a critical factor because burning is rare and light in this fuel-poor, high-elevation environment; at more heavily vegetated, lower elevations, however, lack of recent burning might contribute to tree invasions of meadows (Plate 13), including those in Yosemite Valley.

The climate hypothesis focuses on the dry weather of the 1920s and 1930s, documented in Weather Bureau data. Lending support to this theory is the spatial observation that meadows are in moist parts of the landscape (especially valley bottoms) compared to forests, and the temporal observation that the present meadows developed in response to moist climates of the last 2,500 years (before that time, the climate seems to have been too dry for extensive meadow formation; see Wood 1975, 1984). Also, the mere commonness of the tree invasion—every meadow, large or small, every grassy patch, seems to have invasive trees—argues for a comparably widespread and uniform cause, and climate is such a cause. If this climate hypothesis is correct, the change might be interpreted as cyclical, with the trees destined to die when wetter conditions return. Seemingly contrary to the climate explanation, however, is the fact that California during the twentieth century as a whole seems to have been wet, and uniquely so over the last four centuries, in spite of the dry decades in the 1920s and 1930s (Fritts, Lofgren, and Gordon 1979).

The grazing explanation seems better suited to the ages of the invasive trees, which did not germinate during the dry decades of the last century, and which did germinate earlier inside Yosemite National Park, where sheep grazing ended about 1905, than outside the park boundary, where sheep grazing continued decades longer (Figure 5). The continued apparent health of the invasive trees, moreover, in spite of wet weather episodes, lends support to the grazing hypothesis. Of course, both climate and grazing might have contributed to the meadow invasions, with

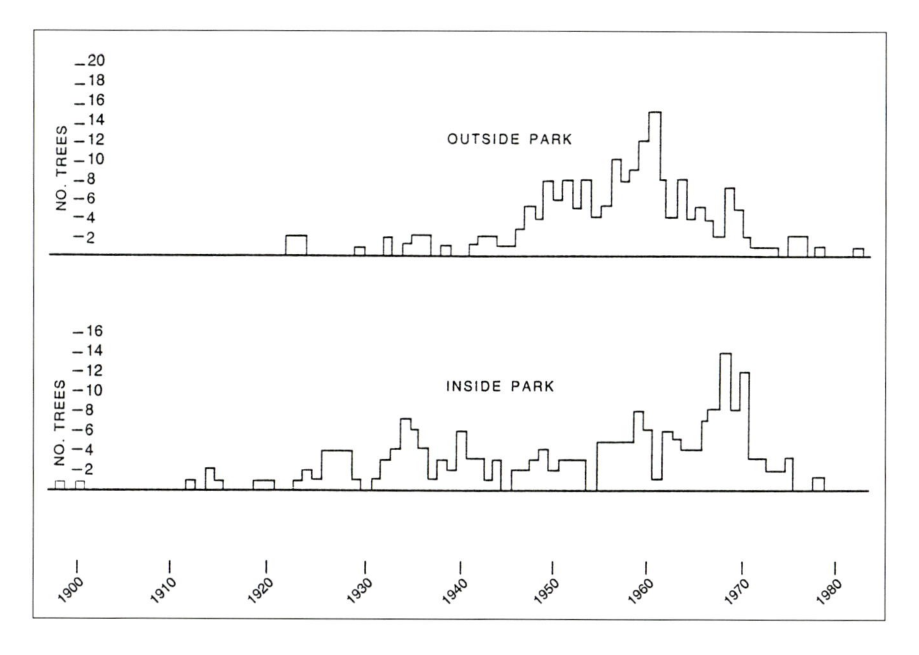

Figure 5. Germination dates of invasive trees in subalpine meadows of the Tuolumne landscape. (SOURCE: Vale 1989)

dry weather creating general conditions suitable for trees and grazing enhancing the opportunities for successful germination.

Almost as common as the tree invasion of meadows but previously not documented in Yosemite (except for Vale 1987), is the thickening of timberline lodgepole forests during this century (Plate 14). In this case, the cause of the vegetation change seems less ambiguous: The warmer climate of the first half of the twentieth century, compared to the cooler conditions of the nineteenth, has made conditions more favorable for trees near their upper elevation limits. For similar reasons, the ground-hugging stands of whitebark pine at still higher elevations, made prostrate by long cold winters, reveal increased upright growth since the turn of this century, likely a reflection of warming (Plates 15 and 16).

Twentieth-century increases in trees in still other environmental situations are more difficult to explain. First, the photo pairs indicate that forests have thickened in small areas within the main lodgepole belt on sites with seemingly good soil and drainage (Plate 17). The question is not so much why forest vegetation occupies these sites today, because they seem highly suitable for lodgepole stands; rather, the intriguing puzzle is why these sites lacked such stands in the past. It is possible that small, localized fires occasionally burned patches of forest and that the thickening represents simply a recovery from such burning.

Second, trees on rock domes and slopes often, although not always, seem more abundant today than earlier in the century (Plate 18). The reasons for such thickening are perplexing. The sparse vegetation on these environmental settings seems inadequate to involve either fire (too little fuel for burns) or grazing (too little forage to attract sheep), and the dryness of the sites argues that increases in trees would be associated with the wetter climatic episodes of the past rather than the drier ones of this century.

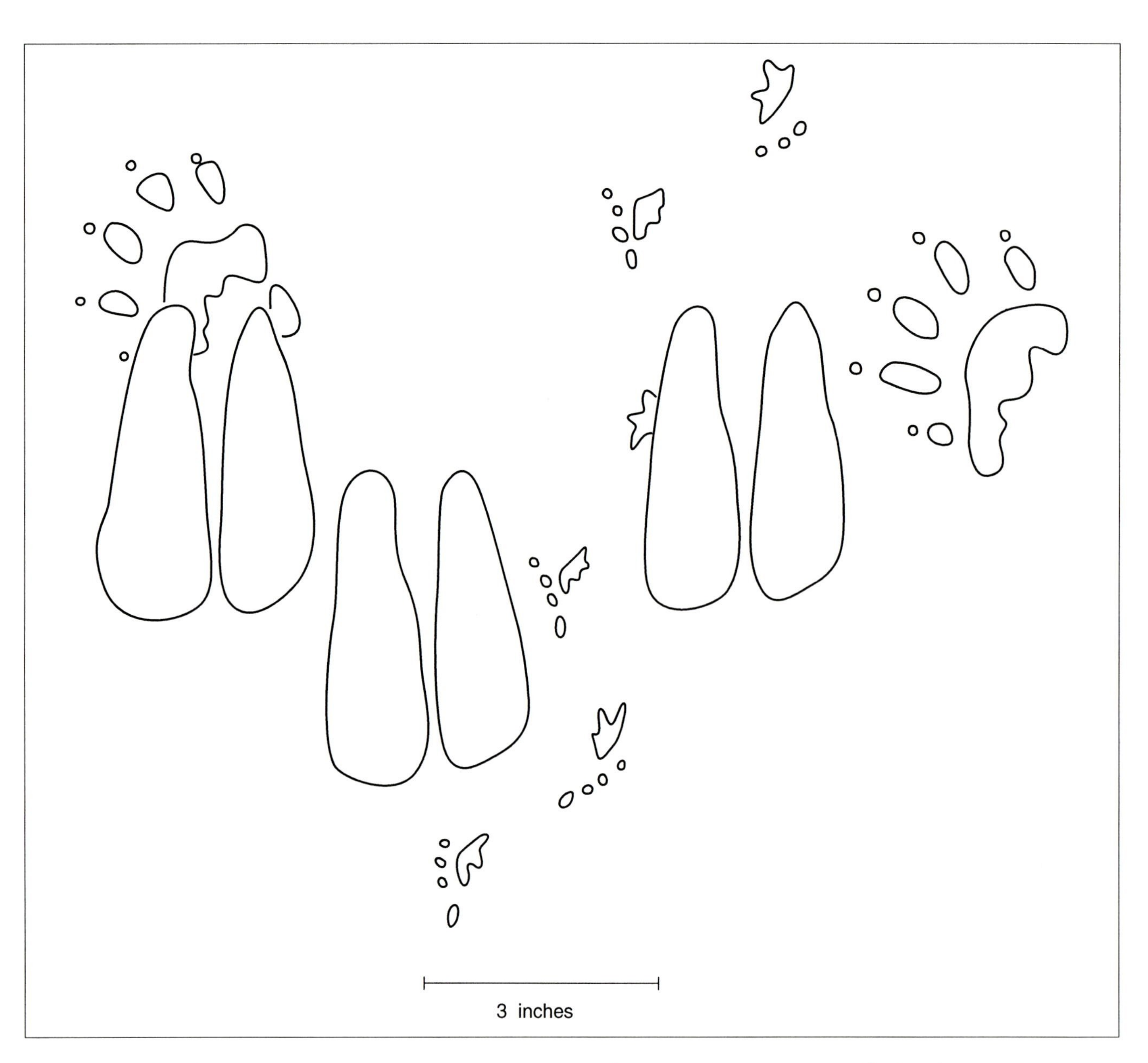

Figure 6. The three large tracks of a bighorn partly obliterate the earlier marks of a toad, which had walked toward the upper right, and those of a wolverine, which had walked toward the left. The frequency with which John Muir saw tracks of certain animals in the wet borders of the Tuolumne River differs sharply from the experience of a modern visitor. Bighorns, for example, have been gone from the Yosemite high country since about 1900, victims of shooting and of competition from domestic sheep. Similarly, the wolverine, largest member of the weasel family, has likely decreased throughout its alpine and subalpine haunts, in part a result of trapping. Other species have increased: Both mule deer and coyotes are more common now than in pre-European times, and fish have been planted throughout the formerly barren waters of the high Sierra (under natural conditions, fish could not swim upstream any farther than the waterfalls and cascades of the middle elevations). Among birds at lower elevations, Steller's jays are more common, at least around campgrounds, as are brown-headed cowbirds; the increase in cowbirds has had the unfortunate effect of decreasing other songbirds because of the cowbird habit of laying its eggs in the nests of many species of vireos and warblers. The future could see still other animals suffering; for example, the Yosemite toad, known only from the Yosemite high country, reproduces best in years with abundant moisture from the melting of winter snow, and climatic warming could prove deleterious to its survival. For a few species, such as the California grizzly, extinction means that no one will ever be able to see them again; the last known individuals roamed the southern Sierra in the 1920s. For still other species, humans can help to reestablish and encourage their numbers. The reintroduction of bighorns in the Lee Vining Creek drainage in 1986 portends days when we modern visitors will be able to find their tracks in the muddy edges of Parker Pass Creek or in the headwater areas of the Lyell Fork. (SOURCE: Gaines 1977; Sherman and Morton 1984; Sumner and Dixon 1953)

Third, Jeffrey pine have increased during the last century on the relatively dry and hot slopes in the Mono Basin east of the Sierra crest (Plate 19). This increase has occurred during the same time period as the meadow invasion in the high elevations, and yet the climatic conditions which would encourage tree establishment in the two environments are opposite: Wet weather should improve germination of trees in the Mono Basin, whereas dry weather should accelerate tree survival in wet meadows. Increases in trees on both the slick rock of the high elevations and the sagebrush slopes of the eastern Sierra might thus reflect nonclimatic factors. Alternatively, trees may be discouraged on the droughty sites in both the Mono Basin and the high Sierra more by winter cold than by summer drought.

Overall, then, the vegetation of the Tuolumne landscape has changed over the past eighty years, with climatic fluctuation as a recurring probable agent (Plate 20). To the extent that climate is responsible, these changes may be interpreted as parts of decadal or century-long cycles in weather, which cause the vegetation to respond on an alternating short-term basis, but which allow a persistence of the general pattern of the vegetation mosaic on a longer time scale.

Climate, however, does not change only in a cyclical fashion; some climatic change is better described as nonrepeating, or directional. Vegetation responding to this type of climatic change cannot be assumed one day to return to its former condition. (The same is true of alterations in animal populations; see Figure 6.) Should the climatic changes of this century be so viewed? If the "greenhouse warming" that is so much in the news does occur, as most scientists predict, the vegetation of the Tuolumne landscape will change directionally. Warmer winters and reduced snowpacks might mean that the upper-elevation forest limit could advance still higher and that meadows will be invaded completely by trees. It is also possible that the main lodgepole forests could become more open or patchy as a result of greater moisture stress associated with reduced snowfall. Drought-enduring shrubs might replace the herbaceous plants of forest openings and cover the forest floor, as they do today east of the Sierra crest. The resulting vegetation could be unlike anything presently known.

Will we who have become transient but constant denizens of the Tuolumne landscape each summer welcome the changes? Will we, like Muir, interpret the new scene as still another form of "ineffable beauty and harmony"? Or will we mourn the loss of the familiar landscape that we have come to know and love?

Plate 4. View of a rock slope east of Tenaya Lake, with Cathedral Creek and the Tioga Road in the valley below. F. C. Calkins 1913; Vale 1987.

The distant forest of lodgepole pine, both in distribution and density, remains virtually unchanged over almost eighty years. Other aspects of this landscape seem similarly static: the great rock fractures on the far slope (formed during uplift of the Sierra and the subject of geologist Calkins's original photograph), the accumulation of broken talus at the base of the cliff, the scattered trees on the bare rock slope, even the log lying on the foreground rocks. Only an increase in the size of the few trees near the log distinguishes the new photo from the old.

Plate 5. West shore of Upper Cathedral Lake, looking southwest toward Tressider Peak.
G. K. Gilbert 1903; Vale 1984.

The large mature lodgepole pine on the right has retained its overall height and shape, its pattern of clumped foliage and dead branches, and even its skirt of small trees. In contrast, the meadow beyond the distant right shore is now broken by invasive trees. The dark shrubs angling upward in a band from the far side of the lake are rooted in the weathered rock along a major fracture in the granite, beside which rises the line of a cliff. Such fracture-related features are typically revealed by straight lines in the landscape. The snowbanks along the base of the north-facing cliff represent wind-blown snow protected from the summer sun; the larger size of these snowbanks in the recent photo compared to the original might suggest heavier snowfall during the previous winter or simply an earlier time of year. The scene as a whole bespeaks a landscape seemingly hostile to life: trees with wind-damaged tops scattered in the few favorable sites with shelter and a little soil. Yet, the persistence of both individual trees and the general pattern of vegetation suggests a harmony, Muir might say a "gleeful harmony," between environment and living things.

Plate 6. View eastward toward the north flank of Medlicott Dome from a low dome beside the Tioga Road. F. C. Calkins 1913; Vale 1987.

Little in the landscape has changed during this century, except for the forest in the lower foreground. In 1913, these trees were dead but standing, a part of the "ghost forest" of lodgepole pine killed here by needleminer moths from 1900 to 1916. By 1987, these snags had mostly fallen, and a forest of young lodgepole pine had become established. Such a replacement cycle might be a common part of the dynamics of the lodgepole forest of the Tuolumne landscape and of the Sierra generally. Within the time of one cycle of moth infestation and subsequent recovery, the vegetation appears to change, but after one completed cycle or with repeated cycles, the vegetation seems dynamically stable.

Plate 7. Mount Hoffmann from southeast of the parking area for the May Lake trail. G. K. Gilbert 1907; Vale 1988.

Ice-age glaciers swept down this slope and carried away the soil necessary to support a dense forest. Even though the open character of the vegetation has been maintained over this century, individual trees have changed. The small lodgepole pine taking advantage of the crack in the rock in the foreground has grown, but its contemporary deformity may reflect its stressful site. Just behind and to the right of that small tree, two snags and a mature tree have fallen sometime during the last eighty years. The density of the foliage on the lodgepole on the left side of the photo view has apparently increased.

Plate 8. View northwest across a meadow in the upper drainage of Rafferty Creek, due east of Johnson Peak. G. K. Gilbert 1903; Vale 1984.

Invasive trees have become established in this upland meadow, obscuring both the large erratic boulders and the distant dome. The meadow sod also might be denser today than in 1907, due either to annual environmental fluctuations (weather or season of the year) or to longer-term change (cessation of sheep grazing, perhaps, which occurred here only a few years prior to 1907). This site was especially difficult for us to locate because the invasive trees have hidden most of the identifying foreground rocks and horizon landmarks. Note that seedling trees are lacking, a suggestion that the conditions encouraging tree establishment no longer exist.

Plate 9. Northeastern edge of Tuolumne Meadows, below Lembert Dome, looking southwest toward Unicorn and Cathedral peaks. R. B. Dole 1913; Vale 1987.

Seedling lodgepole pine are visible in the distant foreground of the Dole view, and these have grown into the dense thickets of today. In the near foreground, over the past eighty years, more rocks have become exposed in the meadow sod, and the exposure of bare granite has increased. Both of these changes might be explained by erosion of fine-textured materials by running water on the meadow surface. Accelerated erosion might be associated with either climate fluctuations, such as warming and drying of the meadow, or human-induced conditions, such as reduced meadow vigor caused by sheep grazing or altered drainage resulting from road construction. Note the unbroken forest blanketing the relatively deep glacial deposits on the slope beyond the meadow, a dramatic contrast to the scattered trees on the mostly bare rock near Upper Cathedral Lake (Plate 5) or below May Lake (Plate 7).

Plate 10. View to the northwest down Parker Pass Creek from a point near the junction of the Mono Pass and Spillway Lake trails. F. E. Matthes 1917; Vale 1988.

F. E. Matthes took his photograph to illustrate the smooth downstream course of Parker Pass Creek, free of major cascades or waterfalls. By 1988, lodgepole pine had so densely invaded the sloping meadow across the foreground that the stream was no longer visible. Moreover, such invasion has occurred through the entire mile's length of the meadow seen in the 1917 photo. A sturdy sapling now spreads its branches where once Matthes spread the legs of his tripod, prompting us to take our photo slightly closer to his foreground. In the 1988 view, one of the authors (Geraldine) is standing on the dark, notched rock visible at the left side of the Matthes photograph.

Plate 11. View northward across Half Moon Meadow, at the head of Yosemite Creek and along the trail to Ten Lakes Basin, which crosses the saddle in the background. F. E. Matthes 1914; Vale 1987.

Invasive lodgepole pine are especially dense along the southern edge of Half Moon Meadow, where they form the thicket in the contemporary photograph. Pockets of invasive trees also occur along other parts of this meadow's borders as well as near the rocks (visible in the original photo) within the meadow. In general, the trees seem to be growing in the most well drained and driest areas of the meadow. Although masked by the foreground trees, the forest on the slope in the background has thickened greatly since 1914. The 1987 photo is taken from a point that may be a little north of the spot where Matthes stood, a vantage that permits an appreciation of the invasive trees but allows a view of the meadow; note the slightly twisted lodgepole pine on the right edge of both photos.

Plate 12. View southward across Emeric Lake. G. K. Gilbert 1903; Vale 1984.

Along the lake's opposite shore, a line of trees marks the well-drained strip of boulders that, sometime in the past, broke off and slid down the granite slope to accumulate along its base. Such rockfalls probably occur sporadically, when slabs of rock have been loosened by weathering, when moisture is especially abundant, or perhaps when an earthquake triggers the slope failure. Many of the individual mature lodgepole pine growing among the rocks in 1903 have died and fallen, but the line of trees has thickened into a pocket of forest. The wet meadow in the foreground remains open and free of trees or shrubs.

Plate 13. View southwest from the northeast edge of Upper McGurk Meadow. F. C. Calkins 1915; Vale 1988.

Outside our main study area, Upper McGurk Meadow, near the Bridalveil Creek campground along the road to Glacier Point, represents a montane meadow in the forests of fir and pine of the middle elevations of the Sierra. Lodgepole pine have invaded the edges and the rock outcrops of this broad grassy opening, as well as the bordering slopes. The small tree in Calkins's photograph is the largest foreground tree in the contemporary view. The open meadow beyond the trees is a lush expanse of grasses and flowering plants, with vast fields of Queen Anne's Lace blooming during our 1988 visit. Fire is probably more important in the maintenance of these meadows within the mixed conifer belt, with taller and denser herbaceous growth and thus greater fuel loads, than it is in the subalpine meadows of the Tuolumne landscape.

Plate 14. View northwest up Lee Vining Creek from a point on the slope southeast of Tioga Pass Resort. G. K. Gilbert 1907; Vale 1984.

The tall mountain hemlock on the left side of the 1907 photograph has died and fallen, but other trees on the slope below the photo site have grown taller, partially blocking the view of the meadow. Lodgepole pine have invaded the meadow and, more conspicuously, the dry slopes beside it. The high-elevation forest of lodgepole pine, on the nearby rock outcrops and on the distant mountain slopes, has thickened greatly. In the recent photo, the Tioga Road hugs the meadow border on a deep fill and circles beside the buildings of the Tioga Pass Resort, in the center of the photograph.

Plate 15. View southeast toward the peaks near Parker Pass, with Parker Pass itself at the far right. F. E. Matthes 1917; Vale 1988.

The bright, open landscapes in the passes along the Sierra crest are home to the alpine sedges and grasses, to the tiny flowering plants like *Draba*, *Erigeron*, and *Potentilla*. These plants share their perennial habit and structural similarities of low stature and nonweedy form with those of the Arctic. But relatively few of the species that prosper in places like Parker Pass also grow in the high-latitude tundras. About one-half of the species are common to at least some other alpine region in the mountainous West, and one-sixth are endemic to the alpine region of the Sierra Nevada (Billings 1988). In total, the Sierra's alpine flora is related taxonomically more to species in the adjacent lowlands than to the distant Arctic. The evolutionary development of the Sierran alpine flora has required relatively long periods of time, from the perspective of a human lifetime, although the pace of that development is likely irregular, with much change occurring rapidly in short periods of time (Gould 1986). The increased growth of whitebark pine, prostrate and windblown in 1917 but more upright and less sculptured by winter gales in the 1980s, indicates that the physical form of plants may also sometimes respond quickly to changed conditions.

Plate 16. View of the flank of Mount Dana and the mouth of Glacier Canyon from a position near the Tioga Road immediately north of Tioga Pass. G. K. Gilbert 1907; Vale 1984.

In some timberline environments of the American West, a krummholz patch of stunted trees moves with time, as exposure to icy winds kills stems on the windward side and protection from them allows new growth and extension of the stand on the leeward. Here, perhaps because of substrate or topography, the stability of the pattern of krummholz stands of whitebark pine on the flank of Mount Dana suggests that no such process is occurring. In fact, even the warmer climates of this century, which probably explain the increased vigor of the willow beside the quartzite knob in the foreground and the denser forest of lodgepole and whitebark pine on the distant slope, seem not to have affected the pattern of krummholz higher on the mountain. Perhaps these patches become established only irregularly, when a combination of conditions enhances germination and survival: a particular year with abundant seed production; a caching of seeds by Clark's nutcrackers, common at Tioga Pass; a chance escape of the seeds from the foraging of hungry rodents; a favorable period of weather which encourages germination and survival of the new trees (McCaughey and Schmidt 1990; Weaver, Forella, and Dale 1990). A coincidence of temporally varying factors might be necessary to allow establishment of a krummholz patch, even on a site that spatially seems suitable.

Plate 17. View south from the trail to Young Lakes, just south of its crossing of Dingley Creek. G. K. Gilbert 1903; Vale 1985.

This view is from the top of a grade on the trail, the last vantage point from which the traveler bound for Young Lakes has such an extensive vista. The small trees at the bottom of the slope, visible in the 1903 photograph, have grown up to mask the view of Cathedral Peak, Fairview Dome, and other landmarks of the Cathedral Range. This growth could represent the recovery of a small patch of forest disturbed by fire prior to Gilbert's visit, or it might reflect the warmer climates of this century.

Plate 18. View north-northeast across the Tuolumne River, below Tuolumne Meadows. F. C. Calkins 1913; Vale 1987.

Wherever broken rock or glacial deposits have accumulated on this bedrock slope, whether in the ravine to the left, on the gentle recess half-way up the dome in the middle of the view, or behind the dark rock mass to the right, lodgepole pine have readily become established. During this century, several of these stands have thickened, a response to more favor-able conditions for tree germination and survival. Such a change is not perceptible, however, to the scores of hikers who, each summer day, walk beside the Tuolumne River here on their way to or from Glen Aulin.

Plate 19. View eastward toward Mono Craters from a spot east of West Portal at the northeastern edge of the Aeolian Buttes. I. C. Russell 1880s; Vale 1988.

A hot afternoon climb brought us to this fine vantage point, where the volcanic rocks and a crowning shrub in the foreground appear unchanged over the last century. A range fire had recently burned some of the lowland, killing the shrubs and encouraging herbaceous plants. The increase in Jeffrey pine on the distant slopes has occurred during the relatively warm and dry weather of this century, an apparent paradox for moisture-demanding trees in such a desiccated environment. The volcanic mountains themselves were built in a series of eruptions late in the Pleistocene, some as recently as about six hundred years ago. Such activity here is not likely concluded, and some scientists think that this area is the best candidate in the country for explosive, destructive, landscape-transforming eruptions.

Plate 20. View eastward across the northeast end of Tenaya Lake. G. K. Gilbert 1907; Vale 1984.

Both change and continuity in vegetation are illustrated in this photo pair. On the opposite slope of gleaming granite bedrock and accumulated broken talus, some stands of trees have thickened, although others, especially on the most rocky sites, seem unchanged. Below the slope and behind the lakeside beach, the open meadow has been heavily invaded by lodgepole pine, as has the drier meadow border. The foreground continues to support scattered trees, but only two individuals are clearly the same in both views. One of the two, the large western juniper in the bottom left corner, moreover, retains its general form and appearance, though its mass of foliage has grown wider. Perhaps John Muir stood here when he imagined himself free of physical needs and limitations, thereby able to observe the dynamics of the landscape:

> Wish I could live, like these junipers, on sunshine and snow, and stand beside them on the shore of Lake Tenaya for a thousand years. How much I should see, and how delightful it would be! (Muir 1911)

REFLECTION:

On Repeat Photography

In the midst of his first summer in the Sierra, when he was obliged to oversee a herd of sheep and its sheepherder, John Muir yearned for a "good time coming, when money enough will be earned to enable me to go walking where I like [in the Yosemite region] . . . [where] every step and jump . . . is full of fine lessons" (Muir 1987). Our own time spent in the Tuolumne landscape for this rephotography project required us to follow in the footsteps of G. K. Gilbert and François Matthes (we could not go simply where we "liked"), but our journeys, as documented in the photo pairs, were "full of fine lessons."

The use of repeat photography as a means of documenting change is popular among those interested in the appearance of landscapes (Goin, Raymond, and Blesse 1992; Rogers, Malde, and Turner 1984; Vale and Vale 1983), especially involving vegetation (Gruell 1983; Hastings and Turner 1965; Humphrey 1987; Rogers 1982; U.S. Bureau of Land Management 1984; Veblen and Lorenz 1991). The study of vegetation change in Yosemite Valley based on a series of old photographs and published in 1964 (Gibbens and Heady 1964), for example, is a classic of repeat photography work.

A first principle in such works is usually to stand in the exact position of the original photographer, and to duplicate the view of the old photo as closely as possible. In our work on the Tuolumne landscape, we always sought the exact spots where the early photographers stood, a quest that usually required hours of careful observation, purposeful meandering, and frustrating backtracking. Occasions do arise, however, when an exact replication may be less informative than some close approximation, and, accordingly, after finding the precise positions and taking the original repeat photograph for the record, we occasionally shifted our feet to take an additional shot if the resulting view provided more information. A view taken by Matthes in 1917 of a dome southwest of Tenaya Lake illustrates this point (Plate 21a). A precise counterpart of this scene today reveals a wall of young lodgepole pine that blocks the view of the distant dome (Plate 21b). If the purpose of the repeat photography is to investigate vegetation change, this exact duplication is useful. If, however, the purpose were to be a study of landform, perhaps of the persistence of exfoliation sheets on the dome or possible erosion of rock from the dome's crest, then a view from a point slightly higher on the slope, a position which provides a vista over the barrier of trees, would be more helpful than one consistent with the exact replication principle (Plate 21c). An even more radical response might be prompted if the interest were in the environment of the swale between the thicket of lodgepole and the dome; in this case, a photograph taken within the swale provides the relevant view (Plate 21d). In this latter view, we can see that the forest remains open, unlike the thickened stand near Matthes's vantage point, although individual snags differ. Also, the large erratic boulders remain in place. The choice of position is therefore a matter of technique; the *pur-*

pose of the repeat photography should be considered when deciding what is the *best* counterpart for an old photograph.

These considerations arise after the contemporary search for the original photographer's location has been accomplished. We were successful in all of our Tuolumne landscape searches, but in terrain less familiar, we have sometimes had difficulty. An example is the 1915 F. C. Calkins photograph of "Washbourn [sic] Meadow on Pohono Trail just east of Bridal Viel [sic] Creek," which we could not find despite several attempts (Plate 22). Searches in the research library and questions to several Park Service personnel were to no avail. People personally familiar with the area may recognize the meadow readily, as we could for most of the Tuolumne landscape views, but the Washburn Meadow site for us remains a puzzle.

a

Plate 21. View eastward toward dome immediately north of Olmsted Point on the Tioga Road. F. E. Matthes 1917; Vale 1988.

This fine example of exfoliation on a granite dome was the object of Matthes's attention, but a screen of trees obscures the view today. Two additional views, both higher and lower on the slope, were therefore recorded. A pattern of lichen on the large boulder in the foreground allowed us positive site identification. For full discussion, see text.

b

c

d

Plate 22. View of "Washbourn [sic] Meadow, on Pohono Trail just east of Bridal Viel [sic] Creek." F. C. Calkins 1915.

This scene, lush with willow and lodgepole (and probably red fir) but devoid of distant peaks, nearby boulders, and even towering relict snags, defied our powers of site location. See text.

THREE

Rocks and Water

As with vegetation, the general earth processes of the Tuolumne landscape may be described in relatively simple terms. The visual water in the landscape—snow, ice, glaciers, streams, lakes—reflects the climate, especially the snow of winter, and the variations in the water features echo the variability in that snowfall. Water has also molded the earth's surface into the characteristic landforms of the Tuolumne region. Before the onset of glaciers two million years ago, streams had already dissected the rising Sierra Nevada. The effect of the subsequent glaciers, then, was to modify the preexisting valleys and ridges, rather than to create new ones.

From one perspective, the rock features of Yosemite are ever changing. John Muir eloquently mused that it is only on the surface that "the wilderness seems motionless, as if the work of creation were done":

> But in the midst of this outer steadfastness we know there is incessant motion and change. . . . The lakes are lapping their granite shores and wearing them away, and every one of these rills and young rivers is fretting the air into music, and carrying the mountains to the plains. . . . Ice chang[es] to water, lakes to meadows, and mountains to plains. . . . [Just as the landscapes that] we now behold have succeeded those of the pre-glacial age, so they in turn are withering and vanishing to be succeeded by others yet unborn. (Muir 1961)

Yosemite's eternal mountains, then, beneath their "outer steadfastness," seem anything but stable. Most of the rock itself is whitish and grayish granite or granodiorite, once molten magma that rose from the Earth's interior and cooled thousands of feet beneath the surface into huge masses of hard rock. Since solidification, the granitic mass has been slowly exposed by the uplift of the Sierra Nevada and the associated erosion of the overlying rock (the reddish metamorphic rock of Mount Dana and Mount Gibbs is the remnant of such overburden). This erosion continues today. Grains of sand and small rocks are carried downslope in a tumultuous avalanche of winter snow. The spring melting of that snow swells a river channel, and the force of the water rolls boulders downstream along its bed. Even after the spring floods have receded, summer showers strike a rocky outcrop and wash down onto a gentle flat more sand weathered from the lofty granite bedrock. From the cliff rising above the flat, an earthquake triggers the fall of a great slab of hard granite, loosened by decades of hitherto imperceptible rock weathering.

From another perspective, today's serious scientist and casual visitor alike can share Muir's sentiment that with the passage of the glaciers "the work of creat[ing]" appeared complete. Erosion since the end of the Ice Age pales in comparison to the shearing force of water driven within frozen monoliths one-half mile deep, several miles wide, and scores of miles long:

> For, notwithstanding [that] the post-glacial agents—air, rain, frost, rivers, earthquakes, avalanches—have been . . . engraving their own characters over those of the ice, [the glacial characteristics] are so heavily emphasized and enduring they still rise in sublime relief, clear and legible through every after inscription. The streams have traced only shallow wrinkles yet, and avalanche, wind, rain, and melting snow have made blurs and scars, but the change reflected on the face of the landscape is not greater than is made on the face of a mountaineer by a single year of weathering. (Muir 1901)

Muir perceived stability and change as varying with different scales of time (Plate 23).

The glacial landforms that seemed to Muir little modified over the ten thousand years since the retreat of the glacial ice remain, not surprisingly, relatively unchanged over the eighty years separating the comparative photos. During the Ice Age, great tongues of ice originated along the crest of the Sierra, as they did near Parker Pass (Plate 24), across the mountain ridges above Tuolumne (Plate 25), and atop other peaks throughout the high country (Figure 7; Plate 26). In all of these locales, the glacially scoured cirques (the erosional bowls where the glaciers originated) and the serrated ridges and mountains, whose slopes were steepened by the glacial ice, persist as boldly as when the ice disappeared (Figure 8).

The ice flows coalesced in the Tuolumne Meadows area to form a vast frozen sea, and then continued on as separate glaciers down the Tuolumne River and up and over a low rise into the Tenaya Creek drainage (where the Tioga Road currently runs), eventually entering Yosemite Valley beside Half Dome. Wherever the glaciers flowed over fractures and weathered rock, they scraped away the broken detritis, exposing the hard, solid granitic bedrock for which the Yosemite landscape is famous (Figure 9; Plate 27). The rock material so eroded was carried by the moving ice and dumped as moraines, landforms that are present

Figure 7. During the most recent period of glaciation, ice covered most of the Tuolumne landscape. (SOURCE: Alpha, Wahrhaftig, and Huber 1987)

throughout the high country but especially prominent in the canyons on the eastern escarpment of the Sierra (Plate 28). Conspicuous throughout the Tuolumne landscape are the glacially deposited erratic boulders, whether on upland slopes or valley bottoms, appearing as if left behind by the glaciers only yesterday (Plate 29). Where the rock was massively unbroken and thus stubbornly resisted erosion, the ice etched the surface with scratches or gouges (Plate 30), and where meltwater beneath the ice rolled boulders along in torrential eddies, the grinding polished the rock as brightly as did the glacial ice itself (Plate 31). All of these remnants of

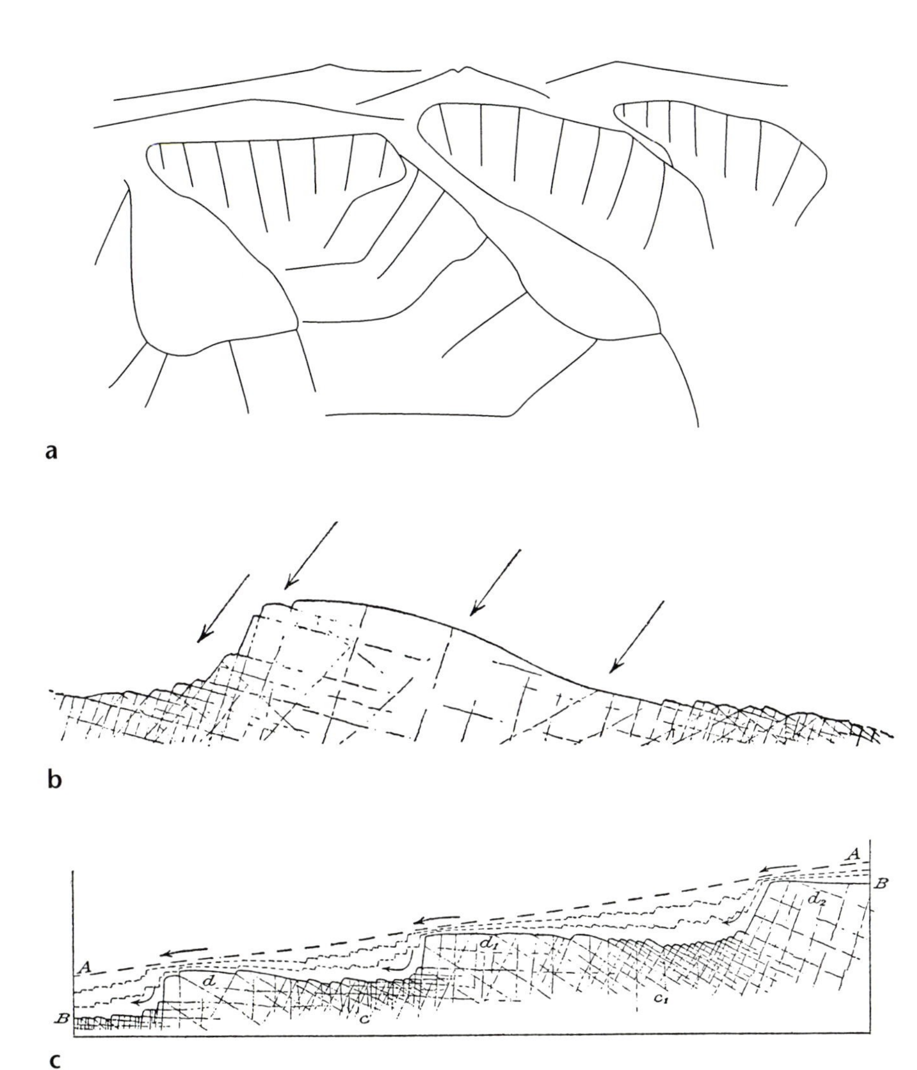

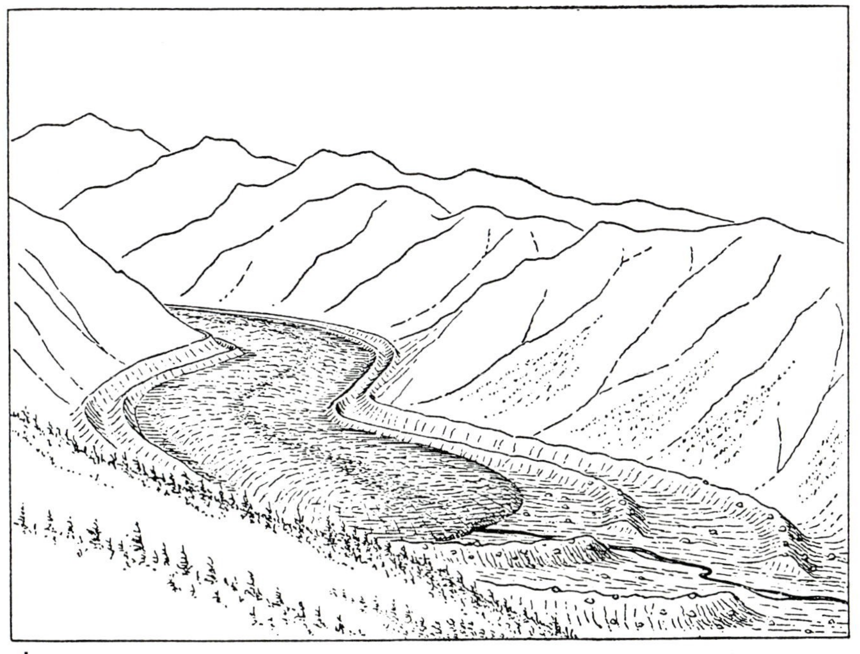

Figure 8. Landscape changes caused by glacial activity. (SOURCES: Figures b, c, and d, F. E. Matthes 1930).

a. Cirques are the steep-sided basins or bowls on mountaintops, such as Glacier Canyon on Mount Dana, or ridges, such as the east side of Kuna Crest, where glaciers originated.

b. Glaciers modified bedrock domes, such as Lembert Dome or Pothole Dome. Ice flowed from right to left in this figure, with the direction of the erosive force indicated by the arrows. The back side, on the right, was polished and scratched; the front side, to the left, was quarried away by the ice.

c. The courses of streams were modified by glaciers. The highest dashed line, A-A, was the bed of the preglacial stream. Glacial ice eroded where the rock was broken by joints, at c and c1, but it rode over the top of less jointed rock, at d, d1, and d2. The resulting irregular bedrock exposure is depicted by the solid line, B-B, and this is the course followed by the postglacial stream.

d. Glacial tongues deposited rock debris called moraines along their flanks and at their snouts in the valleys where they flowed, as in the Lee Vining Creek canyon.

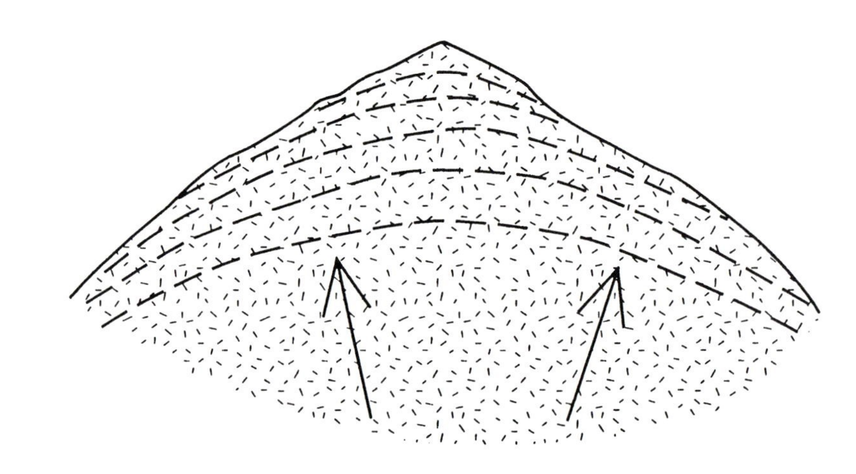

Figure 9. Exfoliation. The domes and rounded bedrock slopes of the Yosemite landscape are not direct products of glaciation. Rather, they have resulted from the erosion of the rock overlying the granite; as a consequence of being freed of its heavy overburden, the granite expands outward. The resulting stress, indicated by the arrows in the figure, breaks the rock into rounded sheets, indicated by the dashed lines. When these sheets fall away, the remaining bedrock is rounded. Glaciation helped to expose the exfoliation features by eroding away the broken rock and revealing the bedrock below, but the fundamental forms were produced before the great Ice Age. (SOURCE: Figure modified from Matthes 1930)

past glaciers appear as fresh today as they did to Muir despite eighty years of weathering and erosion. It is little wonder that people have often considered mountains to be permanent, rising in their tallest majesty since the Earth was born, because they change so little not only within the length of one human lifetime but also over the eons of recorded human history (Figure 10).

The present-day streams, like the Tuolumne River, flowing down the valleys originally cut by rivers in preglacial times and subsequently modified by glacial tongues, can erode little from their bedrock channels so thoroughly scoured by the glaciers (Plate 32). In the meadowy lowlands, the stream channels reveal remarkable spatial stability over eighty years (Plates 33–35). Similarly, the rock slopes that rise above the creeks and rivers seem to have experienced little major erosion. When the glaciers first melted away from these slopes, loose and fractured rock may have fallen as the newly exposed landforms adjusted to the ice-free condition. Yet, such an episode of erosion would have been short lived, and, once adjusted, the slopes have probably shed rock only occasionally, except for the almost ubiquitous loss of minor volumes of sand. In only one of our photo pairs, a view of the face of Mount Fletcher at Vogelsang (Plate 36), could we detect any evidence of a natural rockfall from such a slope.

As much as constancy and stability characterize the rocks over the last eighty years, variability distinguishes the water features of the Tuolumne landscape over all time scales. Both long-term irregularities and short-term cycles characterize these fluctuations (Figures 11 and 12). The concept of cycles in climate—droughts or deluges recurring on a regular basis—are less fashionable today in scientific thought than in years gone by, except for certain regular climatic cycles during the Pleistocene. In the more recent past, irregular sequences of dry and wet, warm and cold, characterize the favored perception. These erratic climatic episodes are superimposed upon the Sierra Nevada's regular, annual accumulation of winter snow, a response to the seasonal southward shift and strengthening of the northern storm track. The winter snowpack melts, moreover, in a regular and repeated cycle each warm season. Therefore, the pattern, but

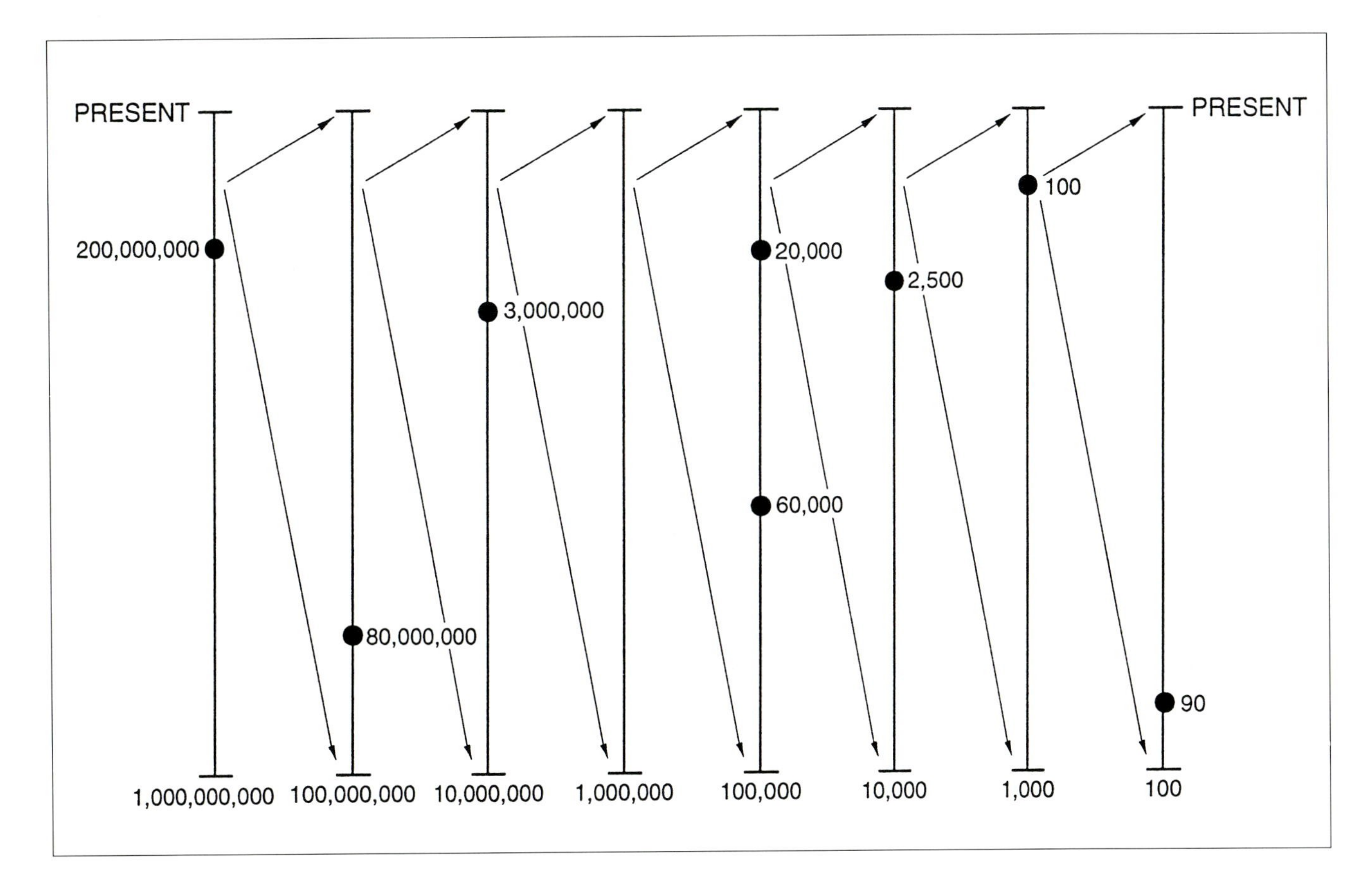

Figure 10. Time Scales and Changes in the Tuolumne Landscape. All numbers in the figure are expressions of years before the present. The granites of the Sierra Nevada were emplaced from 80 to 200 million years ago. The Pleistocene began about 3 million years ago. The most recent glacial advance, responsible for most of the striking glacial features in the Sierra Nevada, started about 60,000 years ago, and its glaciers achieved their maximum extent about 20,000 years ago. Subalpine meadows apparently formed about 2,500 years ago. About 100 years ago, the Tioga Road was built. G. K. Gilbert took many of his photographs of Yosemite about ninety years ago. The telescoping structure of the figure is necessary to depict both geologic and human forces of change on the same graphic. Without such telescoping, the resulting figure would be impossible to read or construct. At a geologic time scale such as that indicated on the far left side of the figure, the century of modern humans on the right side would be less than a millionth of an inch long; at the human scale on the right, the marks separating the present from one billion years ago would be a line over five hundred miles long. A change identifiable on one time scale may be imperceptible at another.

not the size, of a snowbank will tend to be the same early in the summer of a dry year as it is late in the summer of a wet year. The differences in snow patches in two views of a particular scene reflect both the amount of winter snow and the time of year (Plate 24). Taken together, the varying abundance of snowfall in a particular year and the predictable seasonal melting of the winter snowpack determine the water characteristics of a Tuolumne landscape on any given summer day.

When Muir's "winter daisies"—snowflakes—are particularly prolific, they mass to unstable depths on steep slopes, eventually to roar into the valley bottom as avalanches. Avalanches do not occur randomly in the landscape; rather, they tend to recur in certain places conducive to the accumulation of snow. In forested locales, trees may reestablish themselves in the paths cleared by the rivers of snow, perhaps to be battered by subsequent avalanches (Plate 37) or perhaps to mask the event (Plates 38 and 39). Particularly prominent avalanches occurred along the Tioga Road just east of Tenaya Lake and just west of Tioga Pass during the winter of 1986–1987; their paths are easily identifiable near the Tioga Pass entrance station as swaths of broken and bent trees, some of which were carried well out onto the plain of Dana Meadow.

The small Yosemite glaciers which adorn some of the cirques formed during the Ice Age are not remnants of the larger ice tongues of that time. Rather, they have been reborn as new glaciers in response to the relatively cooler and wetter conditions of the last two thousand years. (It is likely that the Sierra Nevada had no glaciers during the relatively warm and dry period after the Ice Age.) At the turn of this century, Sierran glaciers were noticeably larger than they are today (Plate 40), a reflection of the cool period from 1400 to 1900, often referred to as the Little Ice Age.

As snow and ice cover fluctuates on both short-term time scales and an annual cycle, so do the levels of rivers and lakes vary (Plates 41–43). Levels tend to be highest in late spring and early summer, coincident with the peak snowmelt, and then gradually decline into fall and early winter. The smallest ponds and rivulets dry up completely. Summer thundershowers dampen the slopes, freshen the meadows, and, when heavy, increase the flows of the smaller streams, but they usually add little water to the major creeks and rivers (Figure 13).

John Muir was scientifically accurate when he described the rocks of Yosemite as being in "incessant motion and change." Yet, the long time scale of his vision defies reflection in the eighty years separating the comparative photographs, or even in the ten thousand years of post–Ice Age erosion. The Sierra Nevada has risen over the last 15 to 25 million years of intermittent fault movements which have pushed up the range faster than it could erode; it continues to so rise, periodically and rapidly only on the geologic time scale, at the rate of about an inch every century. Eventually, erosion will consume the mountains, as Muir envisioned, in a slow, directional change toward less relief, but the consumption will be so slow that, even if it were to start tomorrow, no one would ever notice.

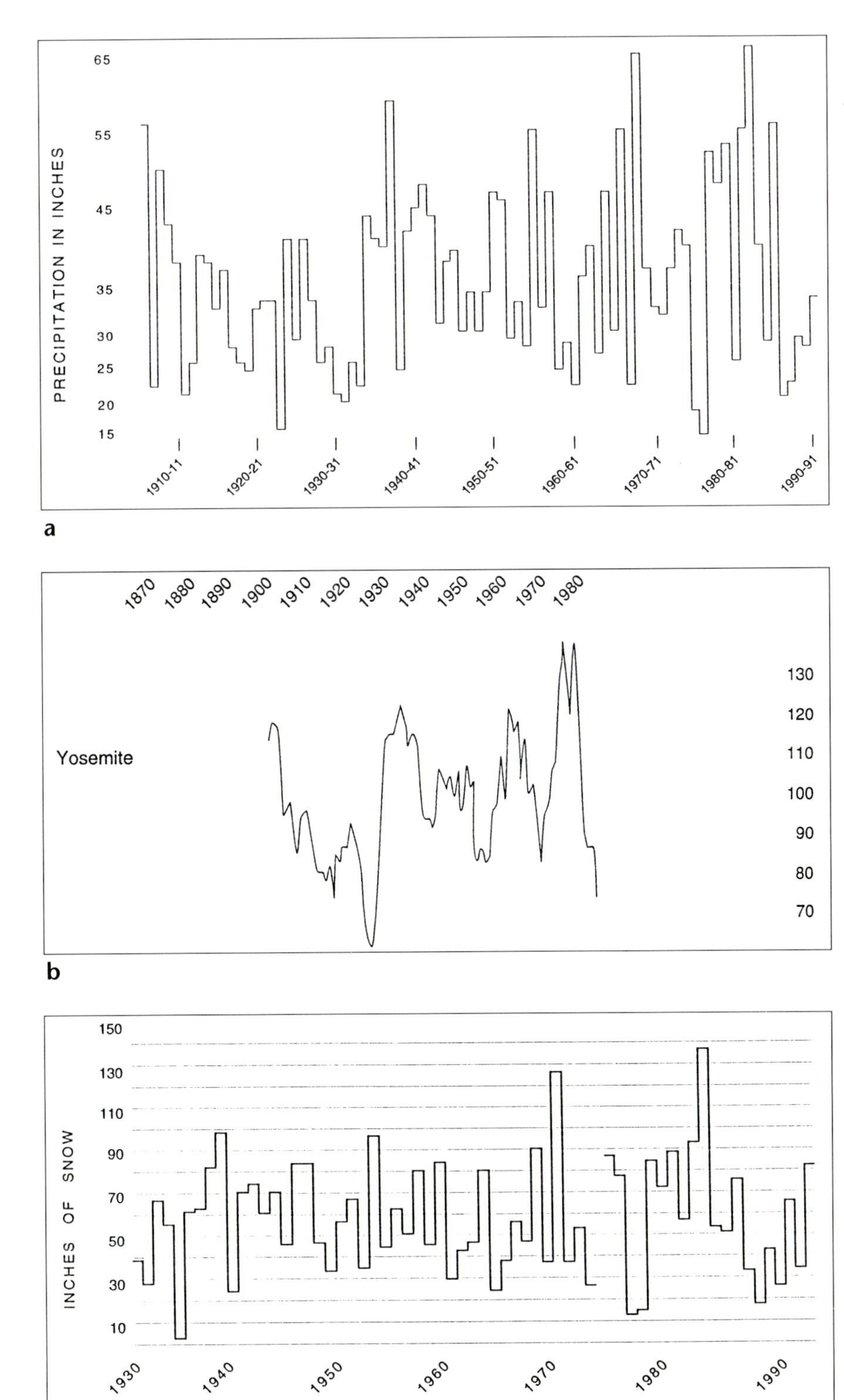

Figure 11. Precipitation in the Tuolumne Landscape. a. Annual precipitation, in inches, in Yosemite Valley, by water year (July 1–June 30). (Complete weather records are not maintained for Tuolumne Meadows; generalizations must be inferred from this closest data set.) The mean, or average, for the period of record is 35 inches, but variation from the mean is normal. Note that dry and wet years are sometimes clustered: Fifteen of the seventeen years from 1917–1918 to 1933–1934 were below the mean, and ten of the twelve years from 1934–1935 to 1945–1946 were above the mean. Still, individual years vary from the group character: The years 1924–1925 and 1926–1927 were wet but within a generally dry period, and 1938–1939 was very dry, even though it was part of a wet episode. After the middle 1950s, the precipitation seems to be highly variable, a pattern noted elsewhere in the world. (SOURCE: U.S. Weather Bureau)

b. A five-year running mean of the percent of annual precipitation (based on precipitation years, July 1–June 30) reduces the year-to-year variability and stresses the patterns produced by clusters of wet and dry years. (SOURCE: U.S. Weather Bureau)

Figure 12. Depth of snowpack at Tuolumne Meadows, end of March or early April (typical time of maximum snow depth), by year. (SOURCE: State of California)

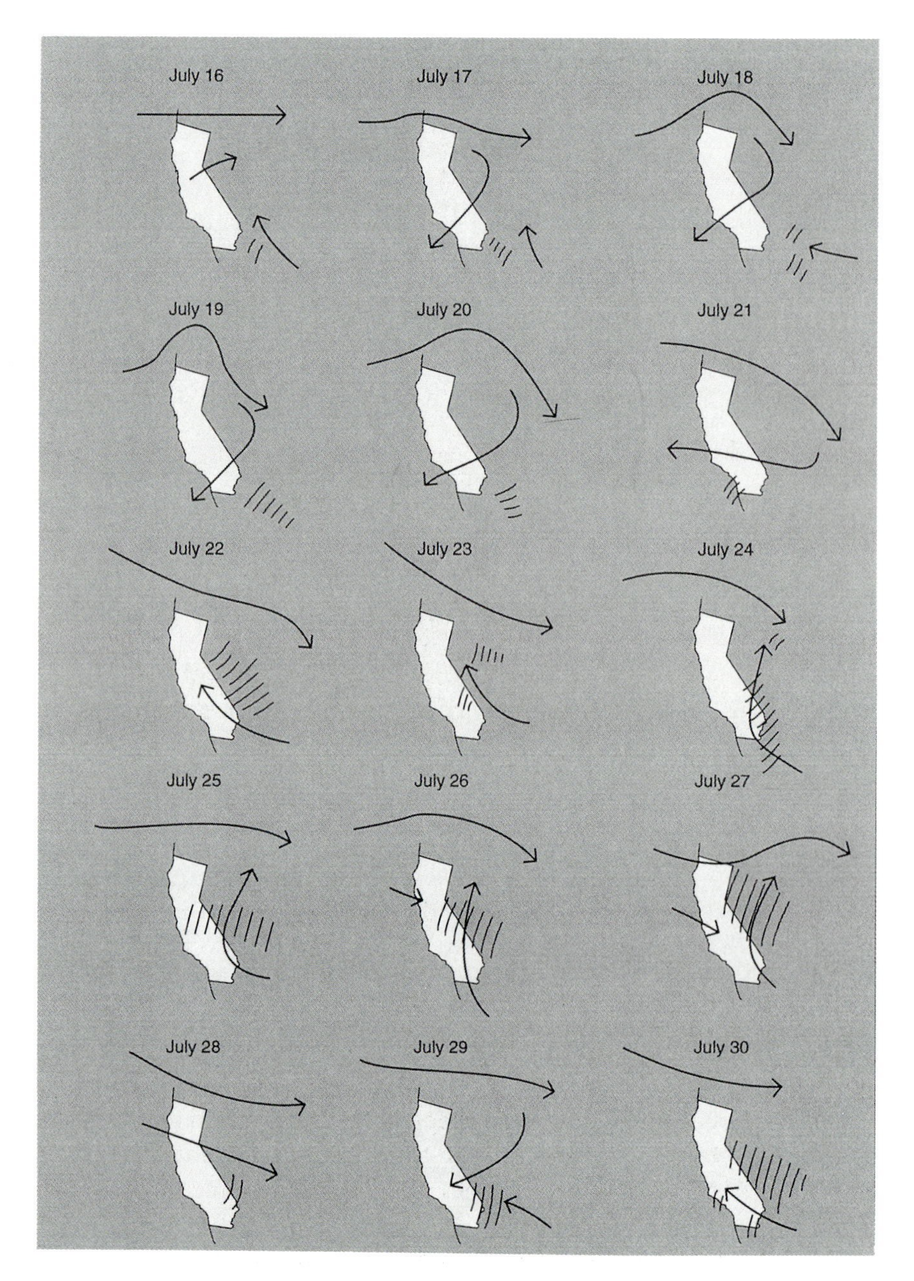

Figure 13. Summer Precipitation in the Tuolumne Landscape. Arrows indicate the generalized winds aloft, 15,000 to 20,000 feet above sea level. Widespread areas of precipitation are indicated by cross-hatching for the days from July 16 to July 30, 1988.

During the summer months, California and the Southwest are beneath winds aloft of little strength and variable flow. When these winds circulate dry air from the west or northeast, as they did from July 16 to July 21, the Sierra is likely to be mostly rainless. When these winds turn around to blow from the south or southeast, as they did from July 22 to July 27, they feed moisture from the tropical Pacific and/or the Gulf of Mexico into the Sierra Nevada, encouraging the development of afternoon and evening showers and thunderstorms. Accordingly, the precipitation records for the rain gauges at Ellery Lake and Yosemite Valley support the generalizations of the weather maps, although the place-to-place variability (the showers are often "hit and miss") means that these particular locales may not receive precipitation when it is falling nearby. From our field study in Yosemite during these days, for example, we know that July 20 was generally dry (and unusually hot) in the Yosemite high country, consistent with the air flow pattern on the map, but that at least some rain fell on the 21 and 22 (when sprinkles occurred at White Wolf late in the afternoon) and that heavy rain drenched Tuolumne on July 23 (in the afternoon, hail closed the road for a time near Olmsted Point, and rain fell from Tuolumne as far west as White Wolf). These wet days, however, were all rainless at Ellery Lake. Contrarily, on July 27, when the rain record indicates a period of afternoon showers beginning at Ellery Lake, Tuolumne and White Wolf were dry, probably reflecting the dry westerly air flow that became established over the west slope of the Sierra Nevada on July 26 (indicated on the maps).

Future changes in the water of the Tuolumne landscape associated with "greenhouse warming," in contrast with vegetation, will more readily be appreciated by even casual visitors—decreasing winter snowfall (Lettenmaier and Gan 1990), disappearing summer snow patches, and shrinking water levels in the lakes and streams during the warm season. On the other hand, in the absence of greenhouse warming, the climate cycles of the late Pleistocene—periods of cold lasting about 100,000 years alternating with much shorter periods of warmth—could mean that the Earth is due for another ice advance.

We who love the present Tuolumne landscape do not perceive the changes in the rocks of the eternally "steadfast" mountains and need the vicissitudes of snow and stream to remind us that nature is indeed in "incessant motion" (Plate 44). Water and vegetation are in flux, but the rocks of the mountains, at least from the perspective of a human lifetime, are indeed eternal.

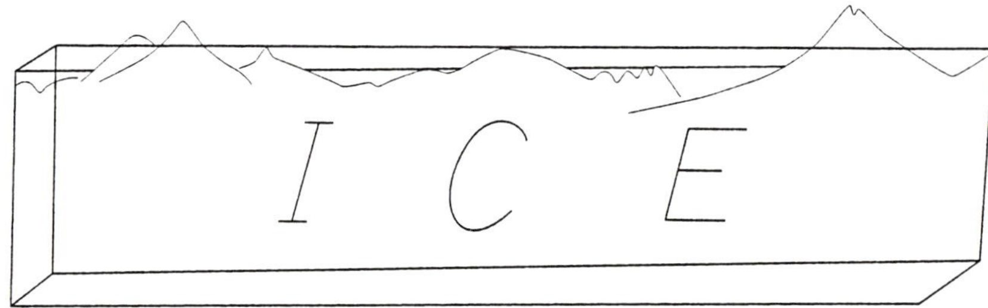

Plate 23. View southward across the western end of Tuolumne Meadows with the Cockscomb on the far left and Cathedral Peak on the far right. F. C. Calkins 1913; Vale 1988.

The Tuolumne ice field was about 2,000 feet deep here, with only the highest several hundred feet of the peaks of the Cathedral Range protruding into the sunlight. (The line drawing suggests the appearance of this glacial landscape.) Whereas the flow of glacial ice tended to smooth and round off the lower slopes in this view, the exposed mountaintops were sculptured by the action of freezing water, which wedged the rock apart, sent fragments cascading onto the ice surface below, and splintered the remaining mountaintops into pinnacled forms. Both change and stability have characterized this landscape: The granitic rock here was emplaced in the subsurface about 86 million years ago, and it has weathered and eroded slowly since; the most recent glacial maximum was fifteen to twenty thousand years ago, but the mountain forms that it produced are largely unchanged since deglaciation; the meadow may have formed two thousand years ago, persisting today as the largest subalpine garden of the Sierra; and, over the last seventy-five years, the small lodgepole pine have grown up along the margin of this garden, gradually converting it to forest.

Plate 24. View southward at Parker Pass, with Kuna Peak on the right and Koip Peak on the left. G. K. Gilbert 1903; Vale 1984.

Pleistocene glaciers originated in the steep-sided, bowl-shaped canyon heads called cirques, which characterize the northeast-facing side of Kuna and Koip crests. Glacial ice flowed out of the cirque on the north side of Kuna Peak and down the canyon that today supports a line of summer snowbanks; it joined the ice field that likely covered the lowland of Parker Pass, and then continued its descent to the left toward the Mono Basin and to the right toward Tuolumne Meadows. The imprint of glaciation on the background ridge is a strong reminder of the power of frozen water, but over the last eighty years little has changed in this alpine landscape, a grand open land of frost-shattered rock, tiny buckwheats, twittering rosy finches, and cool wind.

Plate 25. Ragged Peak from the northeast, near the upper Young Lake. G. K. Gilbert 1903; Vale 1984.

Glacial ice flowed from left to right, filling the lowland in the foreground as high as the base of the vertical cliffs, marked by a linear snowbank on the side of the ridge to the left of center. The sharp break in slope on the upper flank of Ragged Peak to the right also indicates the top of the ice. These landform features persist unchanged during this century, but the whitebark pine have increased in number and density among the glacially deposited boulders in the lowland.

Plate 26. View due west across May Lake to Mount Hoffmann, from the May Lake High Sierra Camp. F. E. Matthes 1914; Vale 1988.

The glacial ice overflowed from the Tuolumne watershed, moved from right to left across the eastern flank of Mount Hoffmann, and continued to the southwest toward Yosemite Valley. The cliff face running across the middle of the photo marks the upper limit of this ice. The castlelike knobs on Mount Hoffmann and the more gentle slope just above the vertical cliff were subjected to the frost-shattering of ice crystals during the Pleistocene but were not overridden by glaciers. May Lake occupies a glacier-scoured basin dammed by a moraine along its southern shore. The grove of timberline trees along the left shoreline has expanded over this century.

Plate 27. View westward across the northeastern headwaters of Yosemite Creek from the summit of Mount Hoffmann. G. K. Gilbert 1907; Vale 1984.

The Hoffmann Glacier originated in a cirque on the north face of Mount Hoffmann, immediately to the right of the photo view. It flowed down the broad valley portrayed here, joined other ice tongues coming from the north (from the right), and continued down the Yosemite Creek drainage to the left. This glacial ice scraped bare the exposed granite slopes. The ice-shattered rocks in the foreground were also produced during the Ice Age when the lower elevations in the view were deeply buried, and only the top of Mount Hoffmann stood above the sea of glaciers. When we climbed Mount Hoffmann in 1984, we shared the summit slopes with Tau Rho Alpha, who was sketching the landscape for the U.S. Geological Survey and whose cartographic skills are apparent in Figure 7.

Plate 28. View southwest, up the canyon of Lee Vining Creek, from a spot on the north valley side, above the Forest Service Ranger Station. F. E. Matthes 1917?; Vale 1988.

The high ridge upon which the photographers stood, and its companion ridge on the opposite side of the valley, are lateral moraines, the accumulations of rock material deposited by a glacier along the sides of the ice tongue. The glacier involved here originated in the headwaters of Lee Vining Creek. During the maximum extent of the Pleistocene ice, this valley was filled with ice to the top of the morainal ridge on the left of the photograph. The ice continued to the left, extending, at one time, into Mono Lake (larger then—and called Lake Russell), where masses of ice probably broke off and floated in the still water. During this century more has changed in the vegetation and human artifacts than in the landforms: The pinyon pine on the ridges and the Jeffrey pine in the valley bottom have increased greatly, and the road and power line (transporting away the electricity generated by the fall of water from Ellery Lake at the head of lower Lee Vining Canyon) are decidedly contemporary.

Plate 29. View northwest across a rock slope immediately west of the campground on the shore of Tenaya Lake. F. E. Matthes 1917; Vale 1988.

The spectacular field of erratic boulders dumped by Ice Age glaciers remains unchanged over this century. Even more remarkably, they probably have changed little since the disappearance of the ice. In spite of ten thousand winter snowpacks and scores of thousands of summer thunderstorms, the boulders suggest persistence, even permanence. This stability reflects the dry granite slope on which they rest. If the same boulders were covered with moist and warm soil, they would decompose chemically much more quickly. Instead, the few which have rolled downslope and out of view, probably pushed by park visitors, are the only hints of change in this century. By contrast, trees have increased in size and number, including one on the spot where Matthes stood, which prevented us from replicating exactly his original view.

Plate 30. View southward across the extreme southwest edge of Tuolumne Meadows, with the east flank of Pothole Dome in the foreground and the Tioga Road at the meadow edge in the background. G. K. Gilbert 1903; Vale 1984.

The chatter marks on the granite slope were gouged out by glacial ice, here thousands of feet deep, as it moved from left to right, up and over the massive rock of Pothole Dome. (These are the glacial features toward which the naturalist-led group is headed in Plate 3.) The mature trees casting the long shadows in Gilbert's photograph are dead snags in 1984, but the line of lodgepole has nonetheless replenished itself and thickened beside the meadow. The dry, light-colored areas in the middle of the meadow appear in both views.

Plate 31. The southwestern end of Tuolumne Meadows, with the west face of Pothole Dome in the foreground and the Tioga Road along the edge of the meadow. G. K. Gilbert 1907; Vale 1984.

These potholes, for which Pothole Dome is named, were formed by water moving beneath glacial ice. The subglacial water apparently flowed upward against this rock face, swirling boulders around in slight depressions, thereby deepening the holes and polishing their surfaces. Potholes are being created today by contemporary rivers, such as the Tuolumne, where they flow over bedrock, indicating that the glacial situation is not a prerequisite to their formation. Trees have increased beside the meadow just beyond the dome, and the forest has become more dense on the slope rising beside the road.

Plate 32. View northwest and downstream on the west side of the Tuolumne River just above White Cascade and Glen Aulin. F. C. Calkins 1913; Vale 1987.

The Tuolumne River is flowing down a bedrock channel exposed by the scouring action of Ice Age glaciers. The present river can do little to alter its bed in such massive rock, and thus it is not much modifying the landscape here today. Logs or other debris may be carried by the high flows of the spring melt, however, to be left temporarily perched on a rock ledge by the receding water levels of summer, as they are in Calkins's photograph, before the next flood again moves them downstream.

Plate 33. View southwest across Lembert Dome, Tuolumne Meadows, and the Cathedral Range from a dome to the northeast of Lembert Dome. G. K. Gilbert 1903; Vale 1984.

Except for some thickening of the trees growing beside the near base of Lembert Dome and for considerable invasion of Tuolumne Meadows by lodgepole pine, this landscape has changed little in eighty years. Even the channel of the Tuolumne River, twisting and untwisting across the meadow, follows the same path. Such stability of the course of the river may be surprising because here it seems free to wander laterally across the unconsolidated rock material underlying the meadow. Perhaps a catastrophic event, a major flood unlike anything experienced this century, is necessary for the stream suddenly to change its channel location.

Plate 34. View eastward across Lyell Canyon at the upstream head of the flat-floored valley (at Lower Lyell Base Camp). G. K. Gilbert 1903; Vale 1984.

The Tuolumne River meanders today as it did in 1903. Trees have increased in number both along the meadow margin and on the opposite slope, although streamside willows may have decreased. The logs lying on the rocks in the foreground of the 1903 photograph are mostly gone today; only one much-decayed log lay among the living trees in 1984. The hikers on the trail, a stretch of the John Muir Trail that links Tuolumne Meadows with Donohue Pass and Thousand Island Lake, suggest the popularity of this route among backpackers, who may have contributed to the loss of the fallen snags by splintering them for firewood.

Plate 35. View northwest across Budd Creek and Tuolumne Meadows toward Pothole Dome from a spot immediately beside the Tioga Road. G. K. Gilbert 1903; Vale 1984.

Budd Creek seems to have straightened its left-to-right path over the last eighty years, perhaps the result of a redirecting of the water flow by the culvert beneath the adjacent Tioga Road. The still mostly unvegetated stream deposit in the foreground of the recent photograph suggests that the creek has continued to disturb the sandy material. The log in the Gilbert photo has fully decomposed (or has been removed by a flood flow of the creek), except for a still-intact segment behind the trees to the right. In the distant background, trees have increased along the base of Pothole Dome.

Plate 36. View south toward Fletcher Peak from a spot north of Fletcher Creek. G. K. Gilbert 1903; Vale 1984.

In a rephotography project, the skyline configuration is usually a crucial clue in finding the precise position of the original camera. An exact match is essential, for a barely perceptible difference in the outline on the horizon can signal that the earlier photographer stood anywhere from a footstep to a ridgetop in another direction. Here, despite our meticulous maneuvering back and forth, up and down, we could not achieve a perfect match of one particular skyline feature without distorting other alignments. We eventually discovered that our problem was the result of a rockfall. Can you find the difference in the skylines? (See the next page for the answer.) This is one of the very few natural changes in the lithology that our photo pairs document, testimony to the stability of the rock forms of the Tuolumne landscape at the time scale of a century. As elsewhere, however, the vegetation reveals more modification, with taller whitebark pine at timberline and perhaps denser, broader patches of sod growing against the rock slabs in the foreground.

Plate 37. View southwest across a meadow to a mountain slope immediately west of the shore of Elizabeth Lake. G. K. Gilbert 1903; Vale 1984.

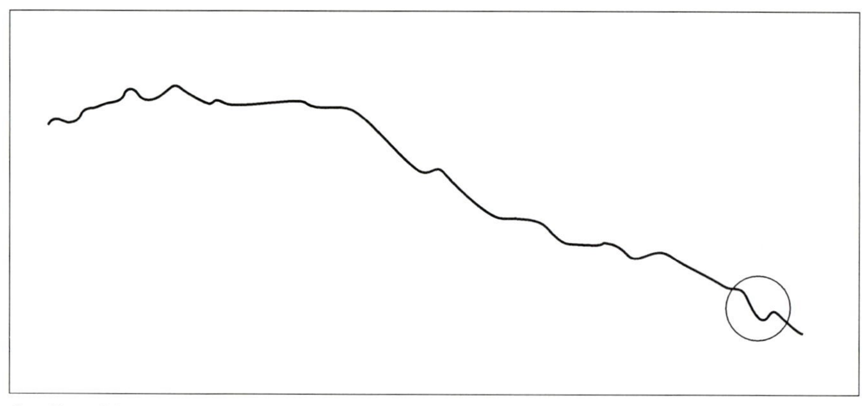

See Plate 36.

Fresh avalanche debris—logs and branches of trees with firm, undecayed wood and intact bark—litter the meadow in both photographs. In the winters preceding both photos, accumulations of snow broke off the east face of Unicorn Peak, above and to the right of the slope in the background, and cascaded down to the meadow. The evidence of avalanching here in two views separated by eighty years illustrates that such phenomena tend to recur in certain favored locales. In 1984, the logs were mountain hemlock, a species that grows on cool, snowy, high-elevation slopes, such as those photographed here. In spite of the proximity of avalanching, trees have increased at the base of the cliff face.

Plate 38. The twin summits of Cathedral Peak, from the south-west near the Sunrise Trail. F. C. Calkins 1913; Vale 1987.

John Muir admired Cathedral Peak more than any other mountain peak in the Tuolumne landscape: "No feature . . . seems more wonderful than the Cathedral itself, a temple displaying Nature's best masonry and sermons in stones" (Muir, 1987). The flanking "avalanche tracks," here revealed by the loss of large mountain hemlock and the deformed stature of the subsequent growth of lodgepole pine, were only confirmation, to Muir, that the mountain was "alive."

Plate 39. View southeast toward Mount Dana from the old Tioga Road between Bennettville and Tioga Lake. G. K. Gilbert 1907; Vale 1987.

In 1907, the small trees broken by avalanching framed, rather than obscured, the view of distant Mount Dana. Visitors to the spot today, with the benefit of this photo pair, could be impressed by the stability here, the persistence of both the mountain and the forest, and the return of people who admire the view or the singing juncoes among the trees. Contrarily, visitors could be struck by change: The damaged lodgepole pine have recovered and matured into their conventional upright form, and the subalpine forest has thickened, perhaps both responses to directional change in climate. In the foreground, the increase in trees has eliminated the low herbaceous plants, and the ground surface is now covered by an organic duff of needles and twigs. Though not revealed by the photos, access has evolved from a long but easy pack trip along the old road from Crane Flat, to a short walk of less than a mile from the new road's broad asphalt pavement.

Plate 40. View southward across Dana Plateau to Dana Glacier and Mount Dana. G. K. Gilbert 1907; Vale 1985.

Dana Glacier hugs the east-facing side of Glacier Canyon and is located, amid the horizontal expanse of snow, at the bottom of the long, sloping rectangular bank of ice that sits in a chute just beyond and below the summit of Mount Dana. The bowl-shaped cirque at the head of the canyon and the open character of the canyon's cross section are both products of the Pleistocene glaciers that nearly filled the canyon and flowed down to the right, although not over the plateau surface. When climates warmed at the end of the Ice Age, this glacial ice probably disappeared completely. Today's glacier has been reborn within the old cirque, a response to the generally cooler conditions of the last several millennia. The warmer conditions of this century have forced a reduction in the size of this modern glacier, so apparent in the photo pair. A singular cold or warm year may leave little imprint in the visual appearance of a glacier; the major changes in glacial ice typically reflect longer-term variability in climate. Responding to the same warming during this century that has reduced Dana Glacier, the krummholz patches in the foreground have increased in upright growth and size.

Plate 41. View northward across the Tuolumne River just downstream from the northwest edge of the main Tuolumne Meadow. F. C. Calkins 1913; Vale 1987.

The level of the Tuolumne River is much lower in the recent photograph than in the 1913 view, a reflection of either less winter snow or a later time of year, or both. The changed pattern of the streamside made our identification of Calkins's position difficult, but so did the increased size and number of lodgepole pine in the foreground, a change repeated in the forest across the river. The nearby logs, so conspicuous in 1913, were almost entirely decomposed by 1987 in this relatively moist setting. Two solitary sandpipers, common birds on shorelines of the Tuolumne landscape, bobbed and called beside the water while we visited this spot in 1987; the same species may have greeted Calkins seventy-four years earlier.

Plate 42. View to the north along the west shore of Harden Lake. F. E. Matthes 1914; Vale 1988.

Harden Lake sits between moraines built by the most recent glacier that filled and flowed down the Tuolumne River Canyon. In this photo view, one of those moraines is the low ridge on the far side of the lake, with the Grand Canyon of the Tuolumne River falling away just beyond. Only a small area, and no stream, drains water into Harden Lake. Thus, the level of the lake drops more during the summer season than a lake with a more dependable inflow. Normal, seasonally cyclical drawdown may help to explain the lower level of the lake in 1988 compared to 1914, but more important is the dry winter and small snowpack of the 1987–1988 winter. Beyond the lake, trees have increased in number, obscuring the rocky moraine that prompted Matthes to take his original photograph. During our visit in 1988, fires burned through the forest a few miles to the south, draping a whitish haze of smoke over the trees and quiet water.

Plate 43. The northwest corner of Lukens Lake, from the east.
F. E. Matthes 1914; Vale 1988.

Lukens Lake is formed by a morainal dam that appears in the photographs as the low forested rise in the background. The lake is fed by a large wet meadow that borders the open water on the east. Even though our photo was taken within a few days of our visit to shrinking Harden Lake, as seen in Plate 42, Lukens Lake seems to be at its normal level. Water stored in the meadow soils and perhaps groundwater seeping from the slopes rising above the meadow, help to keep Lukens Lake at a more constant level than Harden Lake, which is fed mostly by surface water draining its surrounding slopes. Trees have densely invaded the north (right) border of the meadow but not the dense herbaceous growth well within the grassy openings.

Plate 44. View to the east across the Dana Plateau from near the top of the gully leading up from Glacier Canyon. G. K. Gilbert 1907; Vale 1985.

The gentle upland of the Dana Plateau is not only visually unchanged over the eighty years of this century (except for more upright growth of the mostly prostrate whitebark pine) but also topographically unchanged over the several million years of the Pleistocene. Its sloping surface is said to represent the landscape of a time before the Sierra Nevada was uplifted by faulting, dissected by streams, and scoured by glaciers. Even during the height of the glacial advances, the ice tongues flowed around but not over this upland. Ice crystals broke apart rocks on Dana Plateau when the glaciers nearly surrounded it, and, more recently, trickles of water have moved fine sandy materials down the shallow open gullies that drain the plateau's surface. But perhaps no landform, no area of the high Tuolumne landscape, has changed so little for so long as the Dana Plateau.

REFLECTION:

A Landscape Panorama

Plates 45–54. A 360-degree panorama from the southern end of the ridge separating Dana Meadows and the Gaylor Lakes basin. G. K. Gilbert 1903; Vale 1984.

The importance of Grove Karl Gilbert, perhaps this country's most important geologist, lies in how he envisioned the work of geological, and more specifically, erosional, forces through time. The most popular concept at the turn of this century stressed evolutionary development of erosional landscapes through repeated stages over long periods of time; young uplifted blocks of the Earth would be attacked by streams and dissected into rugged mountains, then gradually eroded down into gentle hills and finally into flat plains. To "explain" erosional topography required that the landscape be placed into the correct position in this historical sequence. Gilbert, by contrast, saw landforms simply as a result of the continuous interaction between the force of erosion and the resistance of rock. His was a vision based more on ongoing mechanics and physics, less on historical sequences, and thus he set the stage for the modern emphasis on "process" in earth science. In terms of time, Gilbert emphasized a scale more compatible with the human lifetime.

Gilbert's contemporary, John Muir, also could see scientific lessons in the rocks, so perceptively, in fact, that he was the first to articulate the importance of glaciers in the Yosemite Sierra. But Muir usually cast his observations into the long time scale of mountain birth and death, into a history that ends with the Sierra worn down to a gentle surface. His was, in terms of time scale, a traditional perspective.

Another difference also strongly contrasts the perspectives of Gilbert and Muir. As a scientist, Gilbert focused on the effects of mechanical forces upon the rocks, not the effects of the rocks upon the human soul, and he offered his views as evidence of geologic "truth." Muir, who certainly must have climbed this same ridge, maybe even in the same year, would have seen, with Gilbert, evidence of "a wrinkled ocean of ice," but the scenes he experienced would have stirred his emotions as much as they aroused his intellect:

> Benevolent, solemn, fateful, pervaded with divine light, [this] landscape glows like a countenance hallowed in eternal repose. . . . Nearly everything shines from base to summit—the rocks, streams, lakes, glaciers, irised falls . . . forests. . . . Well may the Sierra be called the Range of Light." (Muir 1901)

Almost a century later, we contemplate the passage of time over the same panoramic landscape, observing it with the physical exactitude of Gilbert, feeling it with the spiritual fervor of Muir.

Plate 45. View eastward toward Mount Dana and Mount Gibbs.

In August of 1907, G. K. Gilbert led a small group of friends on an extended outing into the Tuolumne landscape, inviting them to his "house party": "My cellar is the Yosemite Valley, my drawing room the Tuolumne Meadows, my attic Mono Pass, and my staircase the Tioga Road." Gilbert's biographer, Stephen Pyne (1980), describes the trip as idyllic:

> It was a leisurely, somnolent month, with warm days and cool nights, the laughter of old friends, and miles punctuated by frequent stops for instruction and appreciation. Gilbert skillfully cashed in his rich experience as an exploration geologist for the pleasure of his friends.

Since that year, Mount Dana has risen an inch, maybe an inch and one-half. The avalanche paths through the forest on the west side of the mountain remain distinct, as do the familiar summer snow patches on its upper slope. And it is still possible to enjoy the "instruction and appreciation" of a "house party" in the open and airy "attic " along the Sierra crest.

Plate 46. View southeast, up the drainage of Parker Pass Creek and toward the east side of Kuna Crest.

This side of Kuna Crest is sharply dissected, the result of a series of glaciers which originated in cirques along its protected eastern exposure. These glaciers flowed in a great arc following the present-day meadows, first to the east, then northward down Parker Pass Creek (toward the photographers' position), and finally to the west toward Tuolumne, building the moraines readily visible in both photos as long, linear forested spines. The intervening lowlands, now narrowed fingers, have been heavily invaded by lodgepole pine over the past eighty years.

Plate 47. View south toward Mammoth Peak, with the top of Mount Lyell, 13,114 feet above sea level and the highest peak in Yosemite National Park, poking out above the right flank of Mammoth Peak.

In the mid-1950s, the dean of Tuolumne naturalists, Carl Sharsmith, reported finding dwarf paintbrush, pygmy buckwheat, and other alpine plants growing beneath patches of windswept whitebark pine on moraines in Lower Dana Meadows, portrayed in this view. These plants were growing 1,500 feet below their typical habitats, high on Mammoth Peak or mounts Dana and Gibbs, he suggested, because parts of the moraines are exposed to the full force of winter winds as they howl up the open valley; as a result, these areas become small patches of alpine-like environment in the belt of the lodgepole pine forest (Sharsmith 1956). If climate warming and drying are responsible for the extensive meadow invasion and the high-elevation forest thickening (both changes so evident in this photo pair), these outliers of alpine habitat may be less secure today than they were at the turn of the century.

Plate 48. View southwest toward the eastern end of the Cathedral Range.

One day in early September 1869, John Muir forsook his humble responsibilities as overseer of sheep and sheepherder and succumbed to the glories of "climbing, scrambling, sliding" across the terrain on the distant horizon. There he surveyed "a tremendously wild gray wilderness of hacked, shattered crags, ridges, and peaks. . . . [T]he immense round landscape," he reflected, "seems raw and lifeless as a quarry, yet the most charming flowers were found rejoicing in countless nooks and garden-like patches everywhere" (Muir 1987). The sublime landscape seems equally attractive to subsequent climbers, as Vogelsang High Sierra Camp, one of the popular backcountry destinations in the Yosemite high country, lies in the heart of the Cathedral Range, to the right of center in the photographs, just beyond the head of the broad forested valley with a central meadow. On the foreground ridge and slope, we, like Muir, were drawn to "the most charming flowers"—here nodding clumps of whorled penstemon and spreading phlox, tucked between sheets of white granite.

Plate 49. View southwest toward Tuolumne Meadows and the western end of the Cathedral Range, with Cathedral Peak near the center horizon.

Pleistocene ice covered almost all of the landscape in this view. F. E. Matthes may have stood on this point when he imagined the appearance of the Tuolumne ice field: "The gently sloping surface of this vast mass of ice extended dazzling white, scarcely sullied by . . . moraines, [over] an area of 140 square miles" (Matthes 1930). Only the highest peaks of the Cathedral Range across the center of the photographs and the barren ridge of Mount Hoffmann and Tuolumne Peak farther west protruded above the glaciers. Lembert Dome, visible as a white, sloping, and pointed mass of rock in the right center, and Tuolumne Meadows, showing faintly to the right of Lembert Dome, were buried deeply by as much as 1,500 feet of ice. Such an overriding mass challenges the imagination, but even greater effort is needed to envision the rapidity with which the flowing ice blanket formed. Only a few thousand years seem to have separated the relatively warm and ice-free interglacial period, which ended a little more than 120,000 years ago, from the subsequent cold episode, the glaciers of which were already building moraines in the Mono Basin, east of the Sierra crest, between 114,000 and 119,000 years ago (Phillips, et al. 1990). Once established, however, this period of ice and cold persisted, abruptly yielding to increased warmth only about 10,000 years ago.

Plate 50. View west across Moraine Flat and down the Tuolumne River drainage.

Glacial ice deposited a sheet of morainic debris on the upland bench called Moraine Flat, which today supports a dark, expansive stand of lodgepole pine, the largest area of such forest in the Yosemite Sierra. In spite of the thickening of small areas over the past eighty years, the forest as a whole remains unchanged. Stability at this broad scale does not mean that the forest is static; dieback associated with needle-miner moth infestations and the toppling of individual mature trees by strong winds open the forest canopy, allowing seedling establishment by the light-demanding lodgepole pine (Parker 1986).

Plate 51. View northwest across the lower portion of Gaylor Lakes basin.

Among the large granite boulders and slabs at the far edge of the distant meadow, a careful search might turn up the Mount Lyell salamander. Two or three inches long, this amphibian uses its broad, moist, webbed feet and blunt-ended tail to hold itself onto the surfaces of slick rock, wet from melting snow and polished from past glaciers, as it slowly searches for its prey of spiders and insects (Stebbins 1954). The Mount Lyell salamander is one of Yosemite's rarest and least-known animals. Endemic to the Sierra Nevada, its scattered populations probably reflect relicts of a more widespread distribution during glacial periods, when its habitat needs for jumbles of cool, moist rocks were more common. The climatic warming during this century, likely responsible for the thickening of timberline forests in this view, and the potentials of greenhouse warming in future years, may threaten anew the survival of this small denizen of the Tuolumne landscape.

Plate 52. View northwest across Gaylor Lakes basin.

The vegetation pattern in the Gaylor Lakes basin in this view is a mosaic that primarily reflects moisture availability (Klikoff 1965). Lodgepole and whitebark pine, whether upright or krummholz form, seem restricted to relatively moist areas protected from abrasion by wind-driven snow. By contrast, the open, windswept flats and slopes support ground-hugging alpine plants. In the low and moist meadows, these plants remain green and blooming into mid- or late August. On sites where soil moisture is largely exhausted by the middle of summer, particularly across the fields of boulders, the sheets of gravel, and the dry upland meadows, plants complete their annual growth and flowering by mid-July. The apparent climatic warming and drying during this century may have changed the proportions of habitats, and thus the associated plants, by favoring the gravels at the expense of the dry meadows (Klikoff 1965), a possible change not apparent in the photographs. The same warming and drying may also have encouraged an expansion of the stands of trees into the moist meadow sod.

Plate 53. View north along the ridge that separates Gaylor Lakes, to the left, from Dana Meadows and Tioga Pass, to the right.

Will Neely, who with Carl Sharsmith formed the naturalist team at Tuolumne for a time in the 1950s and early 1960s, once sat on a peak in the distance and saw evidence in the landscape of a former valley since wrenched by mountain-building and dissected by erosion. He prodded the Tuolumne visitor to discover the past in "ordinary ground and to make poetry and meaning of it and the earth and its forces," to visualize "valleys that no longer exist, to fill in by imagination, and to turn into meaningful harmony what may be just 'rock piles' and crags" (Neely 1958). In this view, one that conceals canyons and cirques, it takes only a lttle imagination to envision the Sierra crest to be a gently rolling hilly landscape, perhaps like that of the ancestral Sierra, fifteen or twenty million years ago. Another mental effort will allow the foreground ridge of ice-shattered boulders to appear from beneath the wasting sea of ice and snow at the end of the Pleistocene. No mental conjuring is needed, however, to see the increased robustness of the timberline forest and krummholz over the last eighty years, thanks to the photo pair. Mental image and photographic image combine to allow us to perceive the changes through time in this alpine landscape.

Plate 54. View northeast toward Tioga Pass and the west flank of Mount Dana.

Lured one evening by the long shafts of sunlight among the trunks of the large lodgepole pine in the forest, we doffed our weighty gear and hiked to this ridge top from the Tioga Road, scaring up several deer as they grazed the green grasses along the way. A few days later we returned on a sunny morning to take the photos and to enjoy again the sweeping vista. Several rosy finches and a flock of pine grosbeaks were foraging among the rocks and patches of meadow immediately below the crest, and we could believe that G. K. Gilbert was similarly treated to such company eighty years earlier. Gilbert is said to have admired John Muir's prose, perhaps more than he enjoyed Muir as a companion, at least in the city (Pyne 1980). Both scientists, however, would reconstruct in their mind's eye the now-extinct flow of Pleistocene ice from the Parker Pass Creek glacier northward across Tioga Pass, where the glacier was only 150 feet deep and thus barely buried the lowest flank of the irregular knob on the skyline. Both men would also be intrigued with the increase in forest density during this century, as documented in the photo pair. But whereas Gilbert tended to evaluate landscape change as the product of physical "force and resistance" (Chorley and Beckinsale 1980), Muir typically responded with emotion: "The mountain winds, like the dew and rain, sunshine and snow, are measured and bestowed with love."

Figure 14. Maps of changing access in the Tuolumne Landscape.

a. Before European times, the Mono Trail was the major means of reaching or passing through the Tuolumne landscape. (SOURCE: Adapted from a map in Greene 1987)

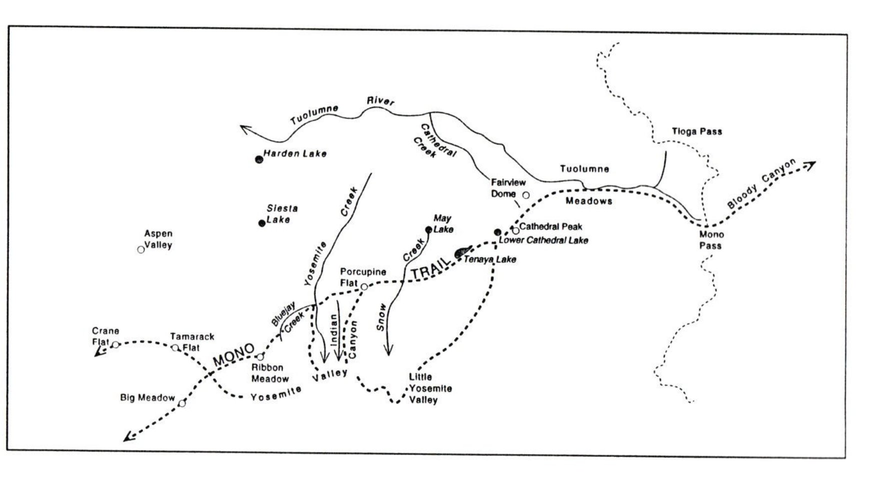

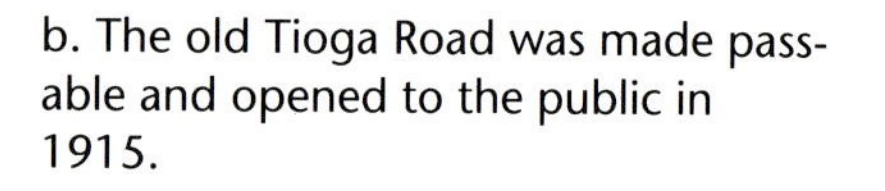

b. The old Tioga Road was made passable and opened to the public in 1915.

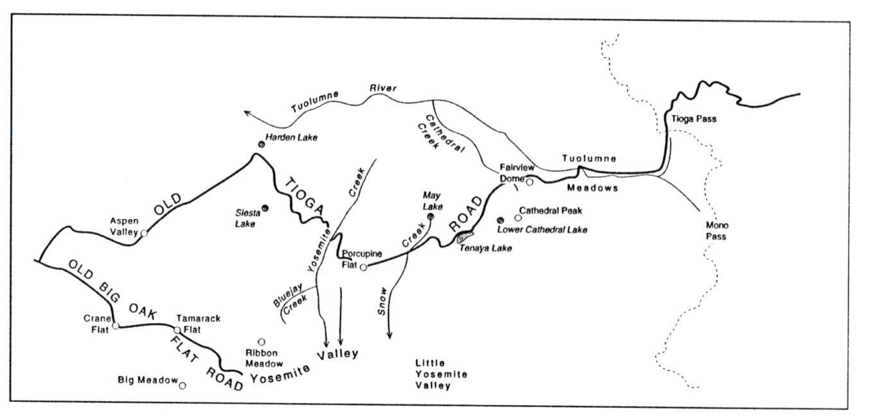

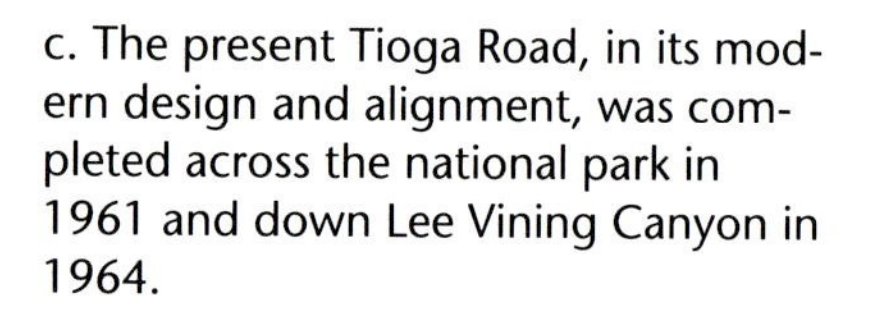

c. The present Tioga Road, in its modern design and alignment, was completed across the national park in 1961 and down Lee Vining Canyon in 1964.

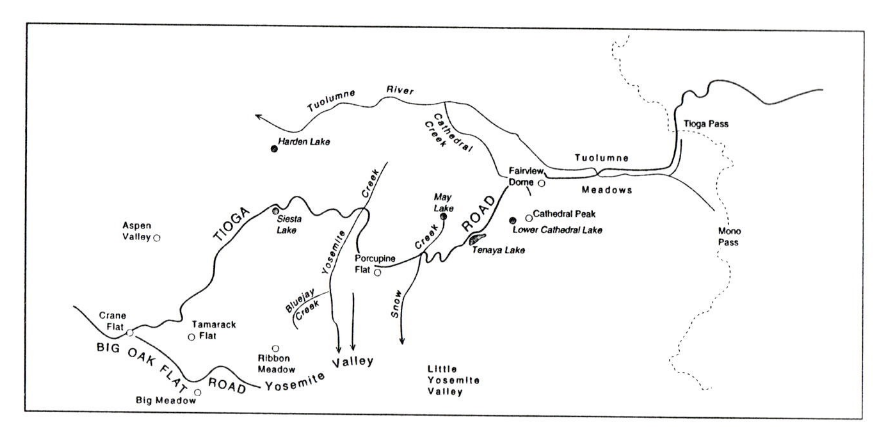

d. The U.S. Army, which administered the national parks prior to the creation of the National Park Service in 1916, recommended public use of the old Tioga Road and the building of additional roads both north and south of Yosemite Valley. The proposed roads depicted in the figure are those that tie into the Tuolumne landscape. The construction of this road system in the early 1900s would have changed forever the environment and the recreational opportunities of the Yosemite high country.

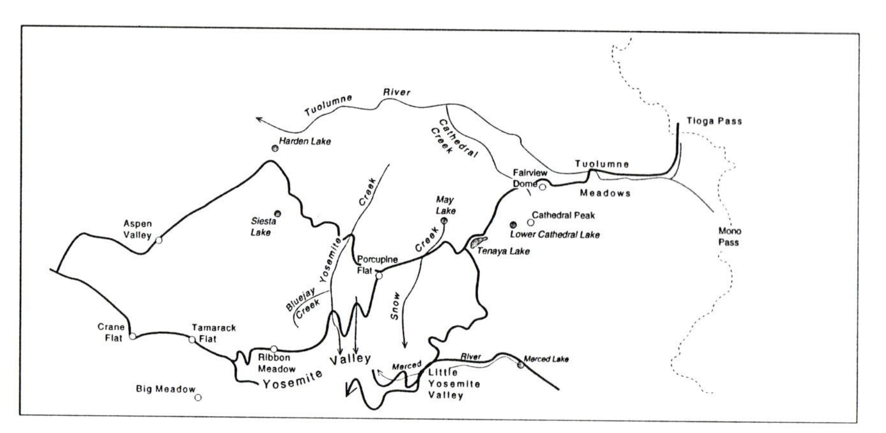

FOUR

Human Impact

> [No experience in the Tuolumne landscape] will be remembered with keener delight than the days spent in sauntering on the broad velvet lawns by the river, sharing the sky with the mountains and trees, gaining something of their strength and peace.
>
> —John Muir, *The Yosemite*

As a haven of solitude, a place where one may "shar[e] the sky with the mountains and trees," the Yosemite high country appeals to the contemplative visitor. Ruggedness of terrain and depth and duration of winter snowpack have long protected this landscape from much of the intense human use visited upon Yosemite Valley below. In past centuries, summer months brought Indians wending their way over the Sierra crest on what came to be known as the Mono Trail (Figure 14). Some came on journeys of trade with burdens of acorns, obsidian, and dried fly larvae, others on hunting expeditions for mule deer and bighorn sheep, which were grazing the lush meadows and browsing the rocky slopes, and perhaps a few were mainly in search of a welcome respite from the sweltering temperatures of the lower elevations. But when winter snows, Indians, deer, and bighorns retreated eastward down Bloody Canyon or westward into the valley and mountain foothills, and the land was subject to only the footfalls of pine martens and black-tailed jackrabbits, the air filled with only the chatter of the chickaree and the howl of the coyote (Plate 55).

With the arrival of white explorers in the mid-nineteenth century came the gradual transformation of the ancient footpaths and environs used a hundred centuries for Indian rendezvous in Tuolumne Meadows (O'Neill 1983). Whereas the Indians' survival depended upon treasures gathered from the Earth's surface, the economy of the newcomers demanded a much firmer wrenching from the high country's blanket of stately rock and grassy forests and meadows. Dreams of wealth from "thundering rich" silver deposits led to the widening of parts of the Mono Trail into the Great Sierra Wagon Road in 1883; eventually to be called the Tioga Road, this route extended from west of Crane Flat to Bennettville (Plates 56 and 57). Miners drove pickaxes into deposits along the Sierra crest at Mono Pass and Bennettville, where they plunged a shaft through 1,800 feet of solid rock. They joined the sheepherders' flocks, already trampling alpine flowers and native grasses in meadow and forest alike. In 1869, John Muir himself, later to become the most prominent defender of the Sierra landscape, spent his first summer in Yosemite, accompanying a sheepherder and his flock of 1,500 from meadow to meadow, where he learned firsthand of the visible changes that could be thoughtlessly wrought by what he called "hooved locusts" (Plate 58). Human use continued to be much heavier in the valley, with its hay crops, dairy cattle, and apple orchards, but one lone individual, John Lembert, homesteaded beside the mineral spring at Soda Springs in Tuolumne Meadows in 1885, fencing off some of his land initially to protect it from the herders of sheep, but then, ironically, renting it subsequently for his own profit as horse pasturage for pack parties of the increasingly numerous visitors.

During these years, one human endeavor entailing a minimum of impact was that of the artist—the writer, the photographer, the painter. Although a pack train of up to twelve animals might be required to haul the equipment of the early photographer or painter, such an entourage hardly matched the impact of the miner or sheepherder. Moreover, it was largely because of the images captured in the paintings of such artists as William Keith and the writings of John Muir that high-country enthusiasts were able to garner public support and lobby successfully for Yosemite's protection as a national park in 1890, a reservation that surrounded the original state-administered Yosemite Grant (created in 1864) until the park became whole in 1906.

But "protection" of the landscape for the enjoyment of, rather than the exploitation by, the public itself requires some measure of human impact. Much as the photographers and painters needed pack trains to gain access to Tuolumne's far-reaching beauty, so also the federal government needed improved, and sometimes new, trails to roust trespassing sheepherders and poachers from the park's boundaries. Under the supervision of the park's first superintendents, the U.S. Army built much of today's trail system. Even after purging the area of economic pursuits, the conflict remained between maintaining the natural scene and providing for recreation. When the Sierra Club was founded by Muir and his fellow enthusiasts in 1892, two years after the establishment of the national park, they naively proposed in their articles of incorporation to "render accessible the mountain regions." One might see the promotion of Yosemite as a place to visit as too successful, for with the increasing number of visitors came a need for improved roads and increased services. Indeed, the Sierra Club, in 1947, found it desirable to strike its own directive to "render accessible" the wilderness.

Movement of visitors within the park continues to create management problems. Although not nearly as destructive as the hooves and maws of sheep or the rifles of poachers, the steady tread of well-intentioned visitors and their pack stock wears rutted footpaths and horse trails, eventually removing the vegetation and contributing to erosion. Such use is now concentrated, with hikers being asked to stay on the main trail whenever possible (Plate 59). Stock animals are limited to trails and may be used only with the accompaniment of a hired guide (Plate 60). The advent of the automobile has brought a more controversial dilemma. Whereas former thoroughfares were primarily only a local eye-catching feature, the cuts and fills and the broad, cleared rights-of-way of modern highways can be seen from vantage points miles away (Plates 61 and 62).

With increasing success at providing easy access has come a growing demand for services to accommodate those travelling the roads and trails. But although our photo pairs document the visual impact of the Park Service's decision to upgrade transportation arteries, they also show evidence of that same agency's refusal to be likewise pressured into offering similar modernization and expansion of lodging in the Tuolumne landscape. No Ahwahnee rises from the sod of Tuolumne. No shuttle bus scurries from parking lot to shops. Instead, the only accommodations are a grouping of one-room, seasonal frame tent cabins. The one permanent building called a "lodge," Parsons Lodge, proudly erected by the Sierra Club in 1915 (the Club had acquired Lembert's homestead in 1912), now no longer caters to the relatively modest lodging needs of the club's members; today, it is staffed by the Park Service for its historical interest only, and most of the memorabilia and documents it housed until a few years ago have been transferred to the Visitor Center. The shanty at the spring itself loses some of its wooden planks with every passing year, it seems, and a beckoning Sierra cup hanging within the shelter no longer entices the tourist to step inside for a cool drink and add the imprint of a modern visitor to the mud and rock floor trod by Muir a century ago (Plate 63). Camping is now prohibited in the meadows (Plate 64), and auto traffic is minimized in the modern campground, across the highway beyond a small convenience store and grill, by requiring that most campers register by Ticketron. Parts of Tuolumne Meadows may be more quiet than they were a century ago, when the meadows were summer home to passing flocks of sheep and Lembert's angora goats; walking in the meadows today, one may reasonably anticipate chancing upon a coyote hunting Belding ground squirrels or a deer grazing at the granite's edge.

Perhaps a more likely place to encounter a Sierra Club member today than at the meadows and Soda Springs is along the circuit of the High Sierra Camps. Any hiker may walk the fifty miles of this circular trail through Yosemite's spectacular peaks and high valleys (mostly in what we consider here to be the Tuolumne landscape) constrained by only the lightest of packs, since the six camps provide tent lodging and prepared group meals. Except for the addition of Sunrise in 1961, new camps have been built only when others have been removed (Plates 65 and 66).

It is easy to visualize not only the appeal of the soothing green coolness of valley and forest but also the challenge of the stark, towering granite to early Indian and later non-Indian visitors alike. Years before rock climbing was in vogue, youngsters must have yielded to the temptation of rolling the biggest boulder over the lip of a steep dome or a precipitous rock ledge, perhaps watching the sparks fly through the darkness of a lampless night, however much the danger to those below. We doubt whether these displacements would be noticed by any park visitor, except as a rock might for no apparent reason come thundering down a slickrock slope from above, and we suspect that the future fate of erratics will not be a major concern in park planning. But for someone engaging in comparative photography, and depending upon geologic features as the most stable parts of the landscape and thereby the most reliable guideposts to location, these missing erratics often lead to considerable frustrated searching and a photograph that requires explanation to the viewer (Plate 67).

The rolling downhill of erratic boulders resembles the building of roads, campgrounds, or lodging. Each represents a directional change in which the exact original condition cannot be easily restored. Nonetheless, efforts have been made to cycle back to a general, if not a precise, natural past—the abandonment of parts of the old Tioga Road, the closing of Smoky Jack, Porcupine Creek, and Tenaya Lake campgrounds, the razing of the Tenaya Lake High Sierra Camp, the decision to allow the old Glacier Point Hotel to rest in the ashes of a 1969 fire. Such attempts to reestablish nature are usually applauded by those who, like us, love not only Yosemite but the national parks more generally (Plate 68). Yet the criticism of developments in the parks by these same nature enthusiasts often seems overstated. Even with roads and a few stores and lodges, the national park landscape is still protected from the myriad of human-initiated changes that characterize areas outside the parks. As crowded and developed as Yosemite Valley may be, it is still a pristine wilderness compared to the surrounding landscapes of dams, mines, agricultural fields, logged forests, and grazed meadows so often abounding immediately beyond our national parks' boundaries (Plates 69 and 70).

More appropriate than to ask whether or not a heavily used national park is "natural" is to question how visitors are affected by their time spent in such a park: What is the quality of today's recreational experience compared to that of eighty years ago? John Muir's turn-of-the-century sojourns through Tuolumne would hardly seem "recreation" by contemporary standards of most Americans (Figure 15). He typically traveled alone, on foot, with a minimum of equipment or supplies—hardly a likely customer for the mall's trendy new Eddie Bauer store or even the Recreational Equipment Company's more traditional catalog. Most importantly, with vigorous intimacy and gentle intensity, he constantly sought an understanding of the nature around him, an understanding that

1939	What to Do and See in Yosemite
Visit the Yosemite Museum	
Take the auto caravan tour of the Valley	
A tour of the Valley in open stages	
Take the loop road	
Take trips afield with a ranger-naturalist	
A 7-day hiking trip through the spectacular high-mountain region	
See the sunrise at Mirror Lake	
Visit the fish hatchery at Happy Isles.	
Camp-fire entertainments	
Outdoor entertainments . . . at Camp Curry.	
See the fire fall	
See the bears fed	

1991	Activities in the *Yosemite Guide*
Valley Visitor Center	
Ranger-led walks	
Interpretive programs	
Tours	
Bicycling	
Horseback riding	
Rock climbing	
Guided backpacking trips	
Art classes	
Outdoor field seminars	
Yosemite Theater	

Figure 15. Visitor activities identified and promoted by the National Park Service, 1939 and 1991. Of the eleven activities identified by the Park Service in 1939, four are no longer possible (auto caravan, fish hatchery, fire fall, bear feeding) and a fifth is not officially encouraged (driving the loop road). In another half-century, will the activities identified in 1991 be similarly changed?

united personal emotion and objective science. Today's visitor crams the vehicle or the backpack with the comforts of home, travels in sixty-mile-an-hour traffic to the busy entrance station, waits in long lines at a park store to purchase a premium-priced carton of granola or bottle of Australian mineral water—all preparatory to embarking on the park experience of competing for prime space at crowded campsite, trailhead, or beach (Plates 71–73). Yet, much of what Muir experienced is still easily available in quiet spots, even close to the roads, and requires few accoutrements. Many visitors, however, seek an "entertaining recreation," rather than the emotional and scientific understanding of nature revered by Muir. Muir neither baked himself on a sunny beach along Lake Tenaya nor wind-sailed across its dark surface (Plate 74). Muir knew nothing of mountain biking or white-water kayaking. He climbed mountains, but not to seek thrills or test his endurance. He hiked trails, but free of the need to prove his worth. (Fox, 1985, quotes Muir as saying, "Hiking is a *vile* word. . . . You should saunter through the Sierra.") He camped and cooked his meals on camp fires, but without the accompaniment of blaring radios, spinning Frisbees, or freewheeling skateboards. His highest "delight" was "sauntering" among the meadows, rivers, domes, and forests. Tuolumne still offers experiences like those that enriched Muir—our failure to find them says more about how we choose to use the landscape than how we have changed it.

Plate 55. View southwest and up Bloody Canyon to Mono Pass from a spot where the national forest boundary intersects the dirt road to Walker Lake. I. C. Russell 1880s; Vale 1988.

Pinyon pine have increased among the shrubs of this arid setting, with one tree completely obscuring the glacial erratic so prominent in Russell's photograph. The lowland ridges extending out and away from the mountain front and sweeping across the width of this view are lateral moraines built when glaciers reached the edge of the Mono Basin from high mountain cirques, whose steep slopes are conspicuous on Mount Lewis in the center. The broad, low dip of Mono Pass makes it easy to see why the Indians favored this route, the Mono Trail, across the crest of the Yosemite Sierra. The easy grades also impressed railroad promoters who considered Mono Pass a likely candidate for the transcontinental tracks. In our 1988 visit, unfettered and vociferous dogs bounding from a nearby rustic abode convinced us to make a circuitous and hasty retreat after taking our photograph.

Plate 56. View east, across Tuolumne Meadows from the Tioga Road where it first enters the meadow lowland from the west. G. K. Gilbert 1907; Vale 1984.

More than the increase in trees at the base of Pothole Dome, on the left, and across the rock slope in the middle distance, the differing methods of transportation revealed by this photo pair invite thought about change, this time in travel. John Muir recommended three days to walk from Yosemite valley to Tuolumne Meadows, climbing Mount Hoffmann on the way, from there to hike to the top of Mount Dana and the saddle between Mount Dana and Mount Gibbs, and finally to return to the valley—a total of more than sixty miles. Such a distance would be relatively easy for the horseback travelers in Gilbert's photograph, and still easier for the contemporary automobile or bus, although neither steed nor sedan could mount the steep, loose slopes of the peaks. Undoubtedly, the increased pace of modern travel prevents appreciation of subtle aspects of the Tuolumne landscape—the delicate flowers of the silvery pussy paws growing in a sandy crevice on Pothole Dome, the flush of a white-crowned sparrow from a willow in the meadows, the sound of water rolling over a gravel bar on the river. Speed, however, does have its assets: an hour of auto travel can take the visitor from quiet dense forests of red fir at Smoky Jack to cool and windy meadows at Tioga Pass to desiccating heat and redolent sagebrush at Mono. Perhaps what is crucial is not the speed of the travel but the sensitivity of the traveler.

Plate 57. View southeast toward Mount Dana with the buildings of Bennettville in the foreground. R. B. Marshall 1898; Vale 1987.

Bennettville, at the end of the original Tioga Road, was developed in 1882 as the main service center for the Tioga Mine. (The major horizontal shaft is to the right, out of the view of the photos.) A bold prediction suggested that the town might grow to 50,000, but the lack of silver led to Bennettville's abandonment only two years after its establishment. The increase in trees on the slopes beyond the town probably reflects the warmer climates of this century, but it could also represent some reforestation after tree cutting by the town's residents. The two buildings obvious in the 1987 view are the settlement's major remains today; "not much of a town left," remarked one man who was wandering about with his wife and daughter while we were there. The buildings may be "not much" numerically, but they add a sense of human history to the Tuolumne region. Though the rough-surfaced, rich reddish-brown exteriors have weathered a century of Sierran winters, the graffiti scrawled on the interior walls and the charred logs and empty cans scattered across the floors portend a less secure future.

Plate 58. View east across Dana Meadows with the summit of Mount Dana just peaking above the mountain flank. G. K. Gilbert 1907; Vale 1984.

Even after commercial grazing was effectively eliminated from Yosemite National Park about 1905, recreational and Park Service grazing continued locally. Today, such use is more tightly regulated and limited to the horses and mules of the federal government and the park concessionaire. Even for these animals, bagged and baled feed from outside the park has largely replaced the forage provided by the meadow grasses from within. As a result, the sight of livestock on the meadows of the Tuolumne landscape is less common than in Gilbert's day. More common, however, are lodgepole pine, which may have invaded Dana Meadows with the cessation of sheep grazing. The treeline forests on Mount Dana have also increased in density during this century.

Plate 59. View southeast, up Lyell Canyon, about one mile north of the Ireland Creek trail, which descends through the valley at the extreme right. G. K. Gilbert 1903; Vale 1984.

Trails are a conspicuous cultural feature in the Tuolumne landscape, even in the backcountry. Increased visitor use sometimes produces deep ruts and multiple parallel paths, although here the trail in the 1980s is better described as simply more continuous and well worn than at the turn of the century. As paths are worn wider and deeper, small and loose rocks emerge out of the glacial-age deposits, to be kicked to the side by hikers, mules, or trail crew workers, as they have been here on the contemporary trailside. The Park Service has decided that the most heavily worn trails through meadows should be relocated into the surrounding forests, where the impacts of hiking boot and mule shoe are visually less conspicuous. But how should we lovers of the Tuolumne landscape react to trails more generally? Should trails be regretted as a human intrusion upon the natural scene, or accepted as a necessary result of visitation? Might they sometimes even be welcomed as enticements to explore the landscape, encouraging visitors to venture beyond the nearby trees or the further rocky outcrop, to explore the willow thicket in the meadow or the rising slopes of the upper valley?

Plate 60. View south to the Cathedral Range from the trail to Young Lakes. C. D. Walcott 1897; Vale 1987.

Growth of trees and movement of rocks make our contemporary photo seem different from Walcott's original view, but the patterns of skyline peaks and of certain large erratic boulders (look especially at the foreground slope to the right of center) convinced us that our vantage point was correct. We sat on the rocks and wandered around on the slope here for several hours admiring the view, observing the forays of a family of mountain bluebirds, and investigating the invasion of the lodgepole pine. The horseback party came by on a short trail ride out of the Tuolumne stables, riding up to this open vantage point for a brief view of the meadows and the peaks of the Cathedral Range. The heavy hooves of the horses, like those of mules, are far more damaging to the tufts of shorthair sedge, more likely to turn up rocks from the thin soil, and more probable initiators of erosion than the lighter feet of humans. Moreover, manure produced by the animals sometimes fouls waters and forms a continuous deposit over long stretches of trail. The continued use of recreational livestock in the Yosemite high country has generated little criticism, surprisingly so, given the strong feelings which other human activities provoke among nature enthusiasts. Perhaps the tradition of horses and mules is sufficiently old that, like the Tioga Road, it seems a part of the Tuolumne landscape.

Plate 61. View northeast toward Tenaya Lake from a high rock on the ridge due south of the May Lake trailhead. G. K. Gilbert 1903; Vale 1984.

Nearly eight million people have traveled the Tioga Road since Gilbert's year. Yet, little seems to have changed in this landscape, except the conspicuous swath through the trees marking the road's right-of-way. This national park landscape is protected from the sorts of drastic alterations that might be associated with such intensive human use elsewhere. No motels line the lake shoreline, nor do private cabins dot the scenic slopes. Even the High Sierra Camp that bordered the far shore for a time in the 1930s is now gone. Contrary to the relatively static appearance of the landscape, the straightness and apparent efficiency of the road itself suggest that the pace of movement through it has increased greatly, thereby decreasing, some would say, the opportunity to appreciate the landscape.

Plate 62. View eastward down the Lee Vining Creek canyon from southeast of the Tioga Road crossing of Warren Creek. R. B. Marshall 1909; Vale 1988.

The old Tioga Road, from Crocker's Station (just west of Hodgdon Meadow), through Aspen Valley and White Wolf, beside Tenaya Lake, across Tuolumne Meadows and Tioga Pass, to Bennettville, was built during the summer of 1883 to allow the hauling of mine machinery to the newly developed Tioga Mine. The road was apparently never used for its intended purpose because the mine was closed the following year. In 1909, when Marshall took his photograph, the state of California was constructing an extension of the road from Tioga Lake to Mono Lake, down Lee Vining Canyon as seen in the photo view. After several years of poor conditions and negotiations over legal rights, the road through the park became the property of the federal government in 1915 (thanks to the effort and generosity of Stephen Mather, first director of the National Park Service), thereby permitting an east-west crossing of Yosemite by car. Only 190 autos entered the Park via Tioga Pass that first year, but subsequently the winding, brake-riding drive through the Tuolumne landscape

has become an ever-popular part of a visit to Yosemite. Had this road construction not occurred when it did, early in the development of the park, it seems unlikely that the present Tioga Road would ever have been built, so strong has the value of roadless wilderness in the national parks become. A wilderness Tuolumne would be different, and some would say better, but we have heard of no proposals to erect barricades to autos at Crane Flat or Lee Vining—although a replacement of the road with a railroad has been suggested.

Plate 63. View east-southeast toward the old cabin at Soda Springs in Tuolumne Meadows. R. B. Dole 1913; Vale 1988.

The walls of the log shelter enclosing the Soda Springs have sagged; the roof has disappeared. The wooden walk raising the feet of passersby from the sticky wet clay is replaced by a modest narrow path through the rivulets, but a low bench invites today's visitor to sit and admire the view. Only a spreading of the meadow trees to the left disturbs the seeming immutability of the natural landscape. John Lembert's homesteading and goat and sheep grazing (1885–1890) and the Sierra Club campground (1912–1973) are only memories here.

Plate 64. View east of the Soda Springs toward Lembert Dome from a low rise in Tuolumne Meadows south of the trail to Glen Aulin. G. K. Gilbert 1903; Vale 1984.

Campsites in and around Soda Springs, especially along the Tuolumne River, were favorites for early park visitors, including John Muir and G. K. Gilbert's scientific party. The Sierra Club, who assumed ownership of the old Lembert homestead, allowed camping for club members until 1973, when it transferred its land to the Park Service. Camping now, of course, is restricted to the familiar campground on the forested slope south of Tuolumne Meadows, where ecological changes caused by humans seem reduced and visual conspicuousness of people clearly lessened. Here, the old campsite appears recovered since Gilbert's day, with increased growth of herbaceous plants and the ever-present invasive lodgepole pine; one tree, in fact, now grows on the exact spot where the woman stands in the 1903 photograph. In contrast, the large pine at the far left of the original view has fallen, with the stump hidden by a small tree. Most strikingly, the growth of trees now hides Lembert Dome, the probable main focus of Gilbert's picture.

Plate 65. View northwest along the beach on the northeast shore of Tenaya Lake. National Park Service 1923–1938; Vale 1989.

The lodge of tent cabins bordering the beach on Tenaya Lake's northeast shore was opened in 1916. Similar lodging was also started at Merced Lake and Tuolumne Meadows that same year, and together they were the first overnight facilities established by a park concessionaire in the Yosemite high country. The staff at each camp included a "fisherman" (in 1990s lingo, "fisher" or "fisherperson"), whose daily catch helped to feed the visitors in the large canvas dining tent. The Tenaya Lake camp served travelers in the Tuolumne landscape until 1938, when it was closed in favor of the present camp at May Lake. The newly created National Park Service in 1916 had encouraged the building of these camps for a reason that seems more reflective of 1990 than of this earlier date: to alleviate crowding in Yosemite Valley.

More than fifty years have passed since the lodge at Tenaya Lake was closed and dismantled. The modern visitor no longer hears the gentle flapping of white canvas in the afternoon breeze or inhales the wispy smoke wafting down from tall chimneys of wood stoves. Instead, one's attention is focused on the natural phenomena beckoning here for millen-

nia—the quiet lapping of water on the beach, the soft call notes of yellow-rumped warblers in the pines. Although the forest to the right may have thickened over the last half-century, the largest lodgepole in the center foreground seems unchanged. Grasses and sedges have increased their cover over the bare sand. A decidedly modern feature of the recent photograph is a group of seminude sunbathers lying on the hot sand beside the log. Sitting beside them is their dog, whose presence here is a violation of the prohibition against pets in areas away from the roads. In contrast to most regulations in Yosemite, which generally have become more restrictive of human activities and more protective of the natural scene, the rules regarding dogs (which may defecate in or near streams, chase wildlife, and threaten humans) have become more lenient. At one time, pets were not allowed overnight in the park at all (and during the day, only in a vehicle), and visitors were required to board their animals at a kennel at El Portal.

Plate 66. View northward up Long Meadow, with Columbia Finger on the horizon. F. E. Matthes 1914; Vale 1988.

Sunrise High Sierra Camp, immediately to the left of the photo view, overlooks the sweep of Long Meadow. The camps, originally built to encourage visitation to the high elevations of Yosemite and to enhance "the Park Service's interpretive responsibilities" in the backcountry, are sometimes criticized today because of "sanitation problems . . . erosion [on trails] . . . and generally negative impact on the area's ecology" (Greene 1987). Yet, the photo views suggest little visual change has occurred here, except for the ubiquitous sweep of invasive lodgepole pine on the foreground sod. Former chief park naturalist and superintendent of Yosemite National Park, Carl P. Russell, commented that "heavy use [of the High Sierra Camps] . . . would ruin their atmosphere" (quoted in Greene 1987). Russell's use of the word "atmosphere," rather than "the area's ecology," may be telling: Perhaps the tossing of Frisbees across Long Meadow by the young employees at Sunrise threatens the atmosphere of the place more than the presence of canvas tent cabins or the strings of mules threaten the area's ecology.

Plate 67. View from Erratic Dome eastward across the lower end of Tuolumne Meadows and Moraine Flat to Mount Dana and Mount Gibbs. G. K. Gilbert 1903; Vale 1985.

About 200 million years ago, the magmas that would become the Sierran granites intruded into, and solidified beneath, already existent rock. Subsequent erosion has stripped away virtually all of this overburden, except for scattered remnants, including the masses of Mount Dana and Mount Gibbs. By imagining an extension of the rock of the two great mountains, both westward over the Tuolumne landscape and upward into the sky, we gain a sense of the volume of rock eroded away to reveal the familiar white granites. Erosion of a much lesser magnitude during this century has swept away most of the large erratic boulders in the foreground, the one remaining being most distant from Gilbert and partially hidden in the recent photo by a swell in the dome's surface. People, who roll the boulders down the slope, are the likely agents of this erosion. Also over the past eighty years trees have invaded the meadows along the Tuolumne River and increased on some of the rock slopes to the left.

Plate 68. View northward across an arm of Tuolumne Meadows toward Pothole Dome from a parking area along the Tioga Road. G. K. Gilbert 1907; Vale 1984.

The invasive trees have been cut out of the meadow here, as in other areas beside the Tioga Road, although those hugging the base of the dome were apparently passed by and allowed to mature. In a national park, such vegetation management might be justified as "nature protection" if the trees have invaded because of meadow drying caused by construction of the Tioga Road, as has been argued. But the distinction between "natural" and "anthropogenic" change is usually unclear: Climate fluctuations, which are possibly influenced by humans, and sheep grazing, which may or may not be a contributing factor to tree establishment, complicate the interpretation of tree invasions as acts of nature or consequences of people. Whether "natural" or "nonnatural," however, tree invasion does directly compete with the attractive stretches of open meadows in the Tuolumne landscape, and some might urge removal of the trees simply in order to preserve the environmental diversity as well as the view. Is Tuolumne a nature preserve free from human interference or a human retreat freed from civilization? Or is it both, with legitimate claims made by advocates of each perspective?

Plate 69. View to the south over Ellery Lake from the slope north of the Tioga Road. G. K. Gilbert 1907; Vale 1984.

Lying two miles below the eastern boundary of Yosemite National Park, Ellery Lake was formed in 1927 when a dam was built on Lee Vining Creek. Water from the lake is sent down a penstock to a power-generating station in the canyon below. The wet meadow, with its willow thickets and twisted stream channel, singing white-crowned sparrows and grazing mule deer, blossoms of shooting stars and lupine, is now just a memory beneath a quiet reservoir. Most visitors admire the lake and probably see it as natural, a desirable part of a spectacular scene. We, who usually regret the landscape change that results from reservoir construction, find that we accept this particular human-created lake, as we do the small impoundment on the Merced River at the junction of the Big Oak Flat and Arch Rock entrance roads. Perhaps our approval here reflects the aesthetics of the view, or perhaps it arises from the "natural" appearance of Ellery Lake, which lacks the usual barren shoreline so typical of reservoirs with less stable water levels. Or perhaps it is simply that the lake has been in existence for as long as we have known the Tuolumne landscape: People seem more ready to accept changes that have occurred in the distant past than changes that take place during their lifetimes.

Plate 70. View westward across the north end of Tioga Lake. G. K. Gilbert 1907; Vale 1984.

The high-elevation stands of lodgepole pine have thickened dramatically over the past seventy-seven years in this landscape just outside the national park boundary at Tioga Pass. More subtle is the larger size of Tioga Lake, whose overflow descends to Ellery Lake (Plate 69). In 1928, two low dams were built to increase the volume of water stored for the generation of electricity at the power station below the reservoir of Ellery Lake. The presence of such utilitarian facilities—in this case for the production of energy—is antithetical to the national park's purpose in protecting the landscape from commercial enterprise. Many people who applaud the prohibitions against such resource development in the parks are equally concerned about the changes associated with recreation and tourism. To anyone who has fought the crowds at Camp Curry or Yosemite Village on a summer afternoon, this criticism may seem justified. But the critique is only partly deserved: Resource development such as dam construction almost always affects the landscape more drastically than the recreation demands in even the most developed national parks. The controversy over O'Shaughnessy, or Hetch Hetchy, Dam (completed in 1923 on the Tuolumne River within Yosemite National Park) was a heated issue because the landscape change it was to bring about (the flooding of Hetch Hetchy Valley) was so dramatic and so apparently irreversible. The dams on Lee Vining Creek are small compared to Hetch Hetchy, but even their modest imprints on the canyon would violate the pristine image of a national park.

Plate 71. View of the east side of Dark Hole, a meadow on the old Tioga Road east of White Wolf just before the road drops down into the valley of Yosemite Creek. F. E. Matthes 1914; Vale 1988.

During this century, the forest has thickened at the far edge of the meadow, and now hides some of the rocks along a moraine built by the glacier that flowed down Yosemite Creek, from left to right, beyond and below the trees in the background. The wet meadow in Dark Hole, however, remains thick and lush. When we took our photograph in 1988, the scene was marred by evidence of an intruder who had driven a car over the soft plants and spongy soil. Crushed grasses, visible at the right side of our view, traced the line of travel, as did patches of destroyed sod, open muddy pools, and large chunks of wood, probably used to improve traction when the car became bogged down in the wetness. A more serious problem than freeing their vehicle befell the joy-riders: Park Service rangers caught the violators, who were fined $500 and made to appear in court. In a national park, individual freedoms are curtailed to insure the preservation of the wild landscape for the majority.

Plate 72. First view of Tenaya Lake. The old photo is the initial view of Tenaya Lake from the west on the old Tioga Road southeast of the May Lake trailhead. F. E. Matthes 1917. The contemporary photo is the initial view from Olmsted Point on the present Tioga Road. Vale 1988.

For the motorist driving the old Tioga Road from the west before 1961, the initial sighting of the Tenaya Lake was one of sapphire, sparkling waters slowly emerging through the treetops. The old road hugged the contour, climbed sharply where the terrain was steep, ran beside or up and over topographic obstructions, and generally conformed to the landscape, providing the traveler with close views of nature. The new road, on the other hand, sacrifices this intimacy by reshaping the contour, imposing gentle grades and sweeping curves upon a rugged topography, and forcing itself through where the landscape resisted. The modern equivalent view of the lake from Olmsted Point suggests to some that the new road is more spectacular, more photogenic, and more effective at showing off the lake's "shining rocks" and the valley's bold vistas, but others might insist that a sense of communion with majestic granite and towering lodgepole has been compromised by a subjection to bleak asphalt and lowering exhaust. Clearly, the personality of the road influences the personality of the experience.

Plate 73. View southward across Tioga Pass, with Kuna Crest in the background. National Park Service 1939; Vale 1988.

Nineteen thirty-nine was a banner year for the Tioga Road—for the first time a drive across the park from Crane Flat to Tioga Pass was "dustless," a ride over a road oiled or paved. Gone, however, was the scrutiny imposed on the first auto travelers: In 1918, a Park Service official would certify the adequacy of brakes and tires, including a spare, and the sufficiency of gasoline before selling the visitor a vehicle permit. Nevertheless, precautions were still necessary, for the middle twenty-one miles remained narrow and winding, with occasional washouts. But today driving the Tioga Road involves the hazards of heavy, fast traffic rather than the inconveniences of slow travel or fallen trees. Still, some experiences remind the visitor that nature is hardly subdued here: the threats of wandering deer (and, since 1986 east of the pass, of bighorn sheep) on the roadway; temporary summer road closings forced by rock slides, thunderstorm torrents, or accumulated hail; winter-long closing because of deep snowpacks. Another sort of persistence here is the continued presence of Ferdinand Castillo, for almost forty years the ranger at Tioga Pass who often exhorts the auto visitor to "watch out for the deer" and usually admonishes the roadside walker to "get off the flowers," but invariably urges everyone to "have a good time."

In 1939, 15,000 vehicles passed between the rock pyramids at Tioga Pass. The annual auto fee was $2.00, comparable to the daily auto permit today; camping, however, was free. A rate of ten cars per hour allowed each vehicle to stop at the pass while its passengers strolled about, perhaps to take a photographic trophy of their mountain-ascending achievement. By contrast, the backup at the kiosk in 1988 often exceeded ten vehicles at any one time. And the photographic urge could not be stilled—periodically a car door opened to eject a video-camera-wielding passenger who shot hastily but happily, all the while keeping up with the sporadically moving auto. Crowded roads and full campgrounds have become commonplace in the Tuolumne landscape, as the accommodation availability sign in the recent photograph indicates. This congestion probably has more effect on the recreational experience than on the biological ecosystem. Such an impact, moreover, is mostly perceptual and definitely localized. It is easily avoided, not necessarily by hiking twenty miles and setting up a backcountry camp, but by leaving the roadside to wander out over Dana Meadows or climb Erratic Dome or investigate a forest of lodgepole pine almost anywhere.

Plate 74. View westward across the northwest end of Tenaya Lake, with the Tioga Road along the far shore. R. B. Marshall 1909; Vale 1987.

We took our photograph of Marshall's view of Tenaya Lake from a point well away from the water's edge in order to illustrate human use of the beach. We doubt that in 1909 much, if any, sunbathing or swimming was a part of the recreational experience in the Tuolumne landscape. Even in the 1950s such activities were much less common than they are today, when water-focused amusements seem for many visitors a primary reason for lingering at Tenaya. Just as striking for its contemporary popularity is the recreational rock-climbing on the dome in the background (and on other domes along the Tioga Road), an activity that usually prompts travelers on the Tioga Road to pull over and watch the climbers clinging to the rock faces or dangling precariously from their ropes. It is unlikely that Muir would have welcomed such an audience for his treks up Fairview Dome or Mount Dana. Indeed, the preponderance of people and the emphasis on apparatus seem to us antithetical to Muir's idea of nature appreciation, as well as to Frederick Law Olmsted's vision of national park purpose: "the development of the contemplative faculty" (Sax 1980).

REFLECTION:

A Walk Along Tenaya Lake

Tenaya Lake has always been a focal point for visitors to the Yosemite high country. Today's travelers typically swerve into the Olmsted Point turnout to frame a first glimpse of its far shores in the lens of a camera or video camera. Many picnic in its waterside stands of lodgepole pine; others sunbathe on the sandy beach or swim and sail in the cold water; most watch precarious rock climbers high on the steep domes beside the road. G. K. Gilbert, in 1907, was especially interested in the shoreline's "rampart of boulders . . . pushed landward by the expansion of the lake ice in winter," and he photographed five views in close proximity along the western shore. His scenes, and one additional photo by F. C. Calkins, together with their contemporary equivalent photos, provide us with a look at time and change of both natural and cultural features of the Tenaya landscape.

Plate 75. View northwest across Tenaya Lake from the southeast bank of Tenaya Creek where it leaves the lake. F. C. Calkins 1913; Vale 1987.

Leaving our car in the parking lot for the Sunrise Trail, we walk beneath lodgepole pine with their ever-present pairs of juncoes, over wet meadow and sandy flat, before crossing the outlet of Tenaya Creek to arrive at this spot on the Creek's east shore which, for Calkins at least, provided a spectacular view of the largest lake in Yosemite National Park. Tenaya Lake was named in 1851 for the chief of the Indians of Yosemite Valley, when an eastward-fleeing band of the tribe was captured here by a military force. The old chief allegedly protested, saying that the lake already had a name, *Py-we'-ack*, "Lake of the Shining Rocks." The domes so conspicuous in Calkins's photograph indeed "shine," with appropriate lighting, from their Pleistocene heritage of glacial polish, but the descriptive metaphor preferred by Chief Tenaya has been lost to signpost and map. Also lost is the bold vista which Calkins saw, a consequence of the maturing of the young lodgepole pine visible in the original view. This pattern of twentieth-century tree growth is common along Tenaya Lake's southwestern shore.

Left of the largest dome, visible on the edge of the old photograph, the dark forest marks the valley of Murphy Creek, descending from Polly Dome Lakes. The creek is named for John L. Murphy, who homesteaded, fenced, and presumably kept animals that grazed the meadows adjacent to Tenaya Lake in 1878. He stocked the lake with trout and built a cabin on its shore near the outlet of the creek. Beginning in the early 1880s, Murphy's Cabin became a meal and lodging stop for travelers, whether tourists seeing the high country, construction crews building the Tioga Road (which hugs the distant shoreline), or wandering scientists and artists such as John Muir discovering new truths and affirming old ones in the tranquil but grand setting.

Plate 76. View southward along the shoreline of Tenaya Lake from a point immediately north of the Tenaya Creek outlet. G. K. Gilbert 1907; Vale 1988.

Rock-hopping across the quiet and shallow water of Tenaya Creek and then wandering northward along the west shore, we come to a rocky point that protrudes into the lake. Turning around and looking back, we have the perspective in these photographs. The contemporary photo replicates the view of Gilbert (note, in both photos, the large rock at the right edge, and the fractured rock on the left side), but the growth of lodgepole pine obscures the vista; even by wedging ourselves through a thicket and crouching beneath a tree, we still could not quite position the camera as Gilbert had, so dense was the band of lodgepole along the shore. These trees obscure the clear view of the boulders, aggregated along the shore by lake ice but initially dumped here by Pleistocene glaciers. Larger accumulations of such morainic debris clog the lowland cut by Tenaya Creek, whose erosional valley side shows to the left, and these moraines form the natural dam that created the lake. Tenaya is the largest lake in Yosemite National Park, but it is hardly unique. Hundreds of lakes dot the Sierran high country, sure evidence of the former presence of the great sheets of ice.

Plate 77. View westward at the extreme northwest corner of Tenaya Lake. G. K. Gilbert 1907; Vale 1984.

Walking around and to the west of the point in the previous view, we come to a large flat rock that borders the lake. The growth of lakeside trees during this century is particularly obvious, as is the line of lodgepole pine in the background abutting the Tioga Road, both old and new alignments of which run beside the lake here. The level of Tenaya Lake seems higher in 1984 than in 1907, perhaps reflecting different times of year. The lake had been recently still higher than the early July day when we visited, as evidenced by the lines of accumulated pine pollen, formerly floating on the water surface, which stain the boulders in the foreground and on the left. The woman standing beside the lake in Gilbert's day is clothed in a long-sleeved dress and broad-brimmed hat, but the contemporary couple is indulging in the mid-century addiction to suntanning, however ill-advisable in this thin atmosphere and reflecting lake and granite.

Plate 78. View southeast along the shore from the western corner of Tenaya Lake. G. K. Gilbert 1907; Vale 1988.

After walking westward along the shoreline, we turn around and look back toward the flat shoreline rock of the previous photo pair. Trees have increased on the distant background slope as they have beside the near shoreline. Willows now mask the line of boulders beside the water. The shallowness of the lake here makes it inviting to swimmers and waders, at least for short periods. Such water-contact activity is particularly modern and not much mentioned by John Muir or other early visitors to the Tuolumne landscape.

Plate 79. View northwest across Tenaya Lake from the western edge of the campground. G. K. Gilbert 1907; Vale 1988.

From the site of the previous photograph, we move back into the trees and look across the lake toward the upstream canyon of Tenaya Creek. Gilbert's picture captures a meadowy shoreline, whereas ours reveals a contemporary young forest. A walk-in campground takes advantage of the cover provided by the trees, where this couple have neatly erected their tent and positioned a stump as a convenient seat beside the fire; the camp table is situated to take advantage of the view without intruding beyond the "camp boundary" post, thoughtfully placed to preserve general public access to the lakeshore. Pitching camp according to one's own perception of the ideal sense of place, once acceptable in a Yosemite of bygone days

when visitors were fewer, is now prohibited even in the backcountry. Studies suggest that the inevitable human alterations of campsites, such as trampling of plants and reduction of organic material on the soil surface, occur with relatively light use, and that increasingly heavy human occupancy adds proportionally less and less to the total change. As a consequence, people are encouraged to reuse existing campsites, thereby concentrating the human-caused environmental change. Problems of maintenance and of heavy human use, however, loomed large here, according to the Park Service: In 1992, it closed this campground.

Plate 80. View eastward across Tenaya Lake from a rock ledge along the southwest shore. G. K. Gilbert 1903; Vale 1984.

Skirting around the western shore of the lake from the previous view, we walk briefly beside the Tioga Road before the shoreline angles to the east, away from the pavement and its stream of cars, and we come to this rock ledge with its fine view of Tenaya Peak. In contrast to the other photo pairs on the lake, here streamside trees do not obscure the vista; the small trees in the foreground crack have succumbed to the stresses of their soil-poor site. Most conspicuous is the loss of erratics, undoubtedly rolled over the ledge and into the lake by people. Modern visitors are concentrated along the near shore of Tenaya Lake, which is the side originally followed by the old Tioga Road, probably because construction crews could easily build a low shelf into the lake along the slickrock shoreline (see Plate 74); this alignment was then retained when the road was rebuilt. The far shore, however, was likely the route for earlier travelers on the Mono Trail, who found the forest easier to walk through than the near shore's "shining rock" slope. Standing here on a warm summer afternoon, we pictured in our mind's eyes the intermittent human stream that had passed along the shores of Tenaya: Yosemite and Mono Indians carrying loads of acorns and

fly pupae, salt and manzanita berries, arrows and obsidian; herders and their dogs coaxing nervous flocks of sheep through the open forest and toward the subalpine gardens of Tuolumne; prospectors with pickaxes and engineers with transits, all dreaming of riches along the Sierra crest; early admirers of the landscape, wending their way among erratic boulders and lodgepole pine on the far shore or over the glacial polish on the near one; Chinese and white laborers, with shovels and blasting powder, cutting into the shining granite beside the water and building a ramp of broken rocks to lift a roadbed up the canyon of Tenaya Creek; park visitors slowly steering their Packards and Studebakers over the oiled surface of the old Tioga Road, with an eye open for the wide places where they might pull off to allow passage of an oncoming vehicle; later park visitors accelerating along the new road while craning their necks out the window to see the rock climbers high on the dome above; and two authors searching for the right spot to replicate this view of granite bench, sapphire lake, and glacial mountain photographed by a geologist nearly a century ago.

John Muir's portrayal of glacial erosion on a rock near Tenaya Lake. (SOURCE: Colby 1960)

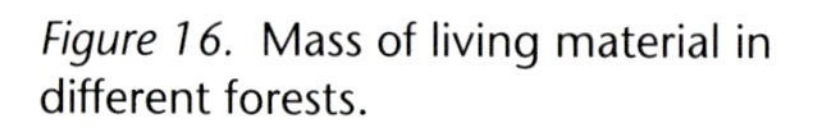

Figure 16. Mass of living material in different forests.

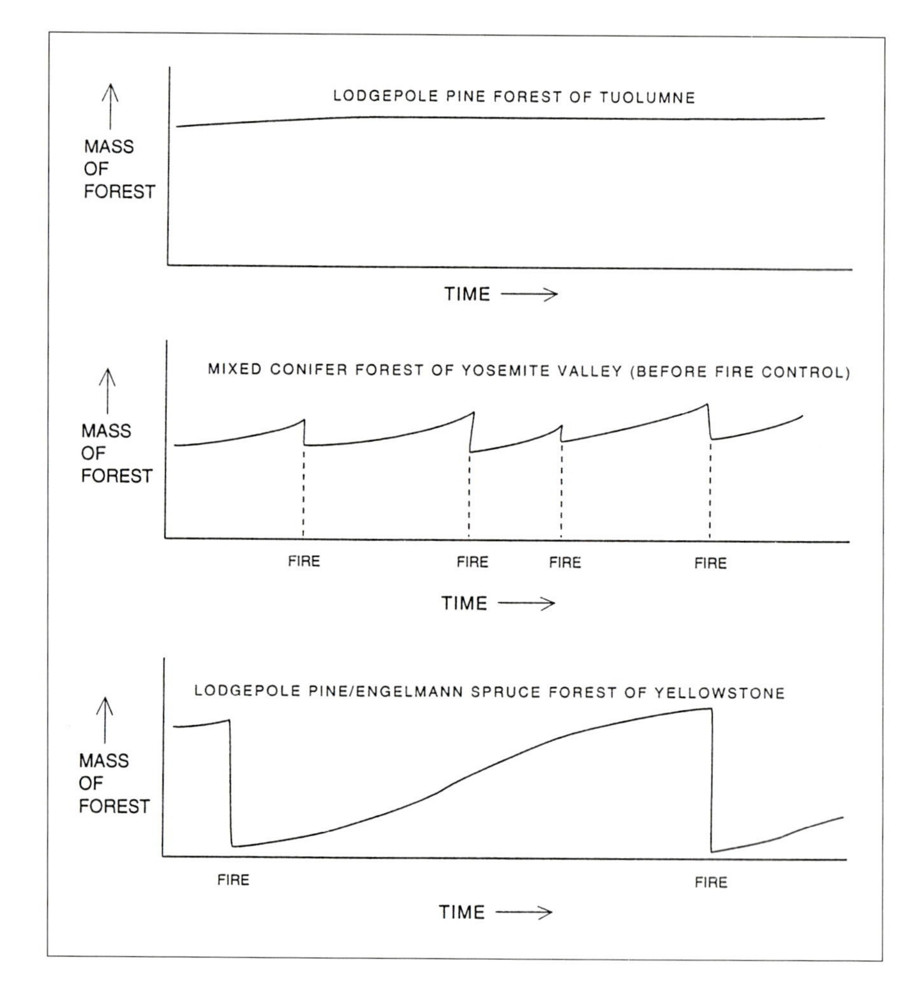

FIVE

Landscape Change Reconsidered

Change characterizes most landscapes. But variation within the pace and pattern of this change is endless. Processes that explain the changes may be widespread; for example, the climatic warming that causes the increase in forest density in the timberline areas in Tuolumne could similarly affect forest density in Colorado or Switzerland. Moreover, the form of changes in a particular environment may be repeated elsewhere; for example, the persistence of glacial landforms in Tuolumne may be similar in form to the persistence of desert landforms in Arizona. Yet, considering all components of the environment—vegetation, soil, streamflow, landform—no two environments change in exactly the same way. The practice of science is concerned, in part, with identifying both the similarities in forms and processes among places and the special or unique features that characterize particular places.

Changes in the elements of the Tuolumne landscape that were studied in this book can be related to various general models of environmental change, models which typify places generally. Specifically, most of the lodgepole pine forests of the Yosemite high country have been stable over time, at least within the limits of the climatic variations of this century (Figure 16). This stability is revealed by the unchanging mass of living material portrayed in the photo pairs. The forest is not static, because individual trees die and new trees germinate and mature, but constant rates of seedling establishment and survival and of tree mortality result in unvarying forest mass. Only the small locales that may be sometimes burned and larger areas that are occasionally subject to outbreaks of needle-miner moths modify this constancy. Other aspects of the vegetation suggest changes that are either cyclical or directional. Tree invasion of meadows, increased density of forests at forestline or on rock slopes, upright growth of krummholz—each might be described as cyclical if responding to cycles in climate or directional if reflecting environmental changes which do not promise a return to a 1900 condition.

Beyond the environs of the Yosemite high country, more conspicuous changes characterize most vegetation types, even if cyclical patterns of variability over short time scales impart stability over long periods of time. For example, the fires which every decade or two burned the lower-elevation forests of pre-European Yosemite, such as those in the valley or below Crane Flat, produced a variable forest condition through time. Fires would sweep along the forest floor, killing small trees where they were growing in the heavy accumulation of needles and branches produced by overarching large trees but skirting the sunny openings where seedlings and saplings were growing without such accumulations. The result was a repeating cycle of biomass accumulation and loss, but within each cycle the amount of variation was small because most of the living mass was in the large trees that were not killed by such fires. An entirely different pattern of change characterizes much of the high-elevation forests of the Rocky Mountains; on the Yellowstone Plateau, for example, fires burn infrequently, perhaps every two or three centuries, but intensely, typically killing most or many large trees. Such a fire regime produces a dramatic cycle of change in forest mass, with the trees increasing in size between the fires.

In contrast to variability which is cyclical in short time scales but stable in long time periods, some vegetation change is abrupt and directional, or noncyclical. For example, Byrne and Edlund (1990) suggest that during the late Pleistocene, a period of major climate change, coniferous forests at middle elevations in the Yosemite Sierra were transformed suddenly by fires, in each locale perhaps by a single burn, into the modern vegetation of pine and oak. Under what conditions might the fires of the 1980s and 1990s in the upper foothills west of Yosemite convert the preburn forest of mixed conifers and oaks into a cover of mostly oaks and other broadleaf trees or shrubs?

In contrast to vegetation covers, which are often influenced by frequent disturbances, notably fires, and sometimes altered by abrupt, directional change, soils are more typically characterized by slow and gradual development. The Yosemite high country's soil, unlike its vegetation, is no exception to the usual. Steadily but extremely slowly, the soils deepen as the granitic rock weathers. Within the soil, materials such as clay or iron may eventually cement other soil particles together into the hard layers that characterize some Arctic forest soils or the famous pygmy forest soil of California's northwest coast (estimated to be at least several million years old). Still another soil characteristic, the increase of organic material in the soils beneath the lodgepole pine of Tuolumne, might reach a steady amount, reflecting a balance between the rate of addition of material by needle- and twig-fall and the rate of decomposition by biological decay.

The glacial landforms of the Tuolumne landscape are mostly unchanged since the end of the Ice Age, thus paralleling over a much longer time scale the constancy of the lodgepole pine forests. However, the glacial features might also be viewed as the products of a pulse or episode of erosion during the time of ice advance, a pulse which was preceded

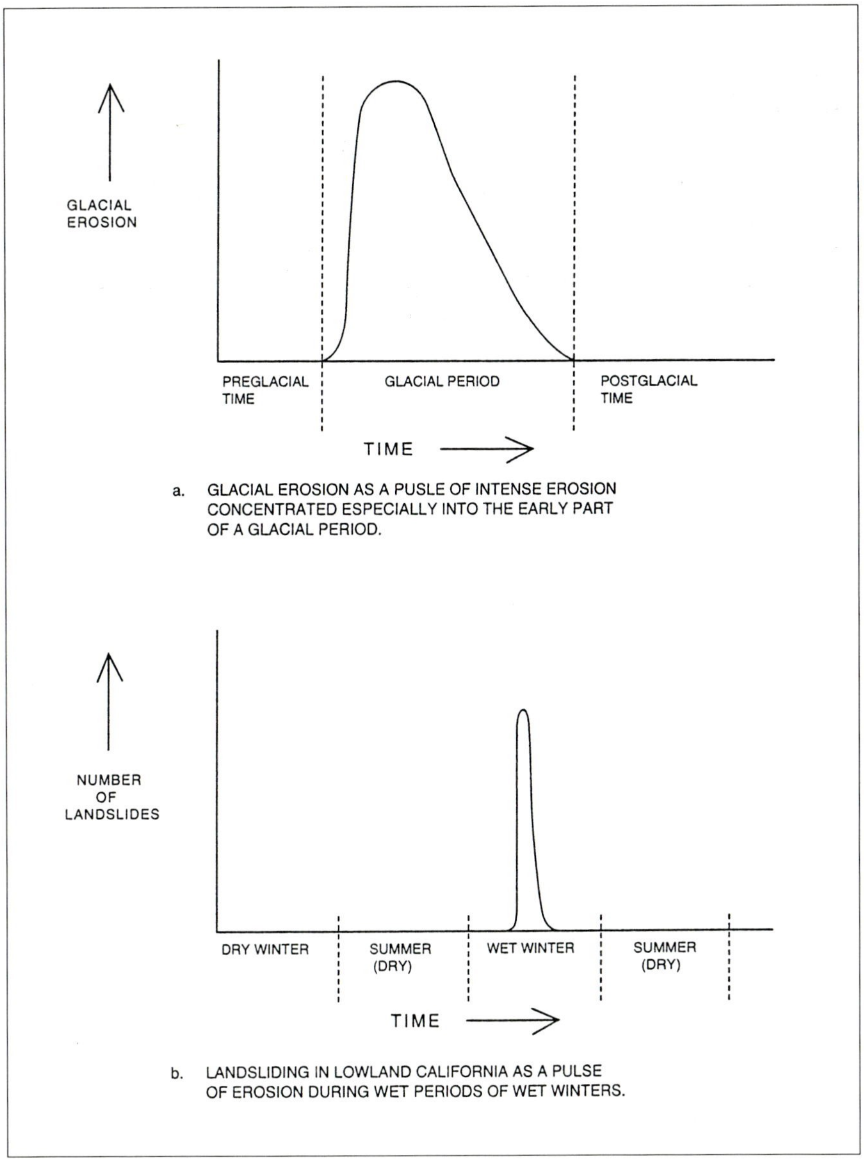

a. GLACIAL EROSION AS A PUSLE OF INTENSE EROSION CONCENTRATED ESPECIALLY INTO THE EARLY PART OF A GLACIAL PERIOD.

b. LANDSLIDING IN LOWLAND CALIFORNIA AS A PULSE OF EROSION DURING WET PERIODS OF WET WINTERS.

Figure 17. Pulses or Episodes of Erosion.

and followed by warmer, ice-free conditions such as those of the post-Pleistocene period (Figure 17). Further, the rate of erosion during the time of ice advance was probably uneven, with erosion concentrated into a still shorter period early in the glacial advance when the landscape was being first assailed by ice and when the most easily erodable rock was available to be entrained by the flowing ice.

This characterization of Tuolumne's glacial landform development as reflecting periods of short but intense erosion would be consistent with many other types of erosional landforms. For example, rockfalls, such as those of Yosemite Valley, which account for most of the erosion on cliffs, are erosional events that are infrequent and periodic, but involve large

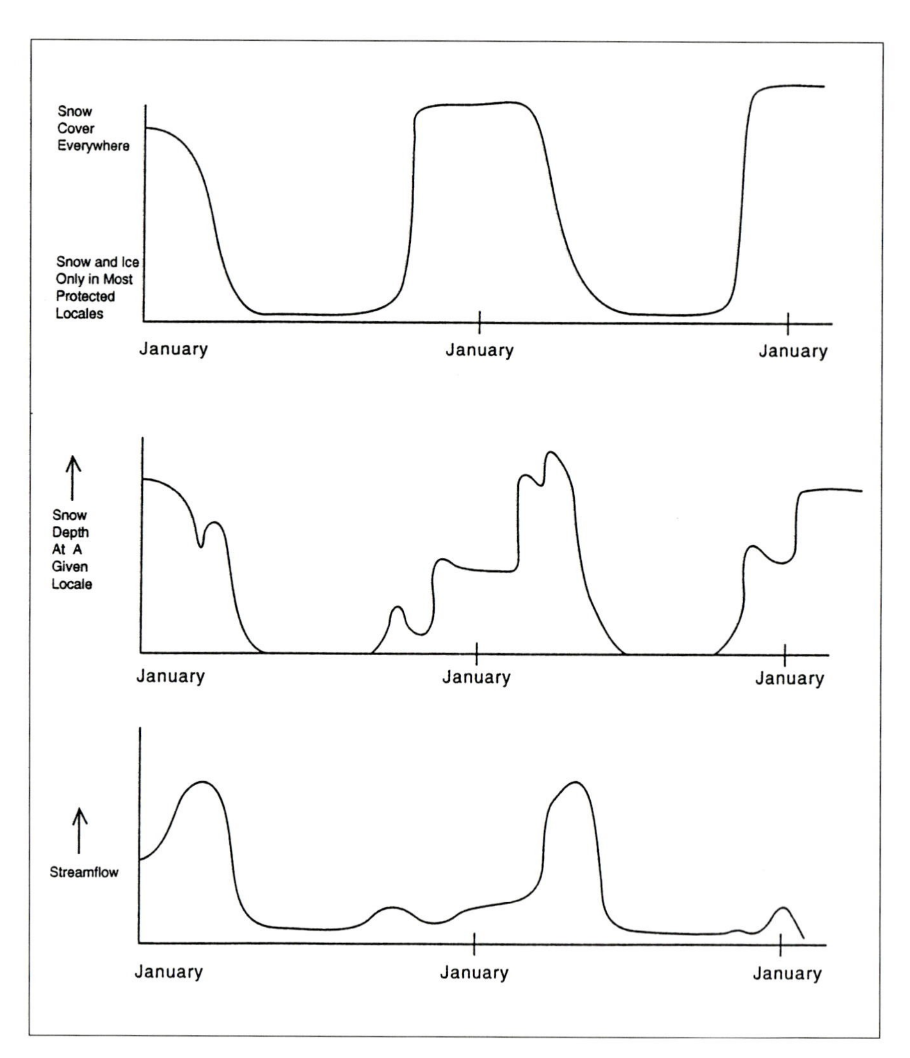

Figure 18. Annual cycles of snow cover, snow depth, and streamflow in the Tuolumne landscape.

masses of rock; in contrast, the frequent and more steady fall of weathered sand grains from the cliffs is less important as an influence on cliff form or development. Similarly, the landslides of earth and soil in the Coast Ranges of lowland California not only occur during a particular season of the year (winter) but also happen most commonly during unusually wet winters and, even, during the wettest times of such winters. In addition, the erosion by streams is greatest during short periods of high waterflow, during which times the bulk of the total erosional work of streams is accomplished. Finally, the erosion by ocean waves is strongly concentrated into short and infrequent periods of storminess and their associated high surf. This model of change—pulses of major change separated by relatively long periods of little change—may thus be common among erosional landforms, and, in fact, has application to other aspects of environments as well.

Together with the yearly growth and flowering of plants, the annual

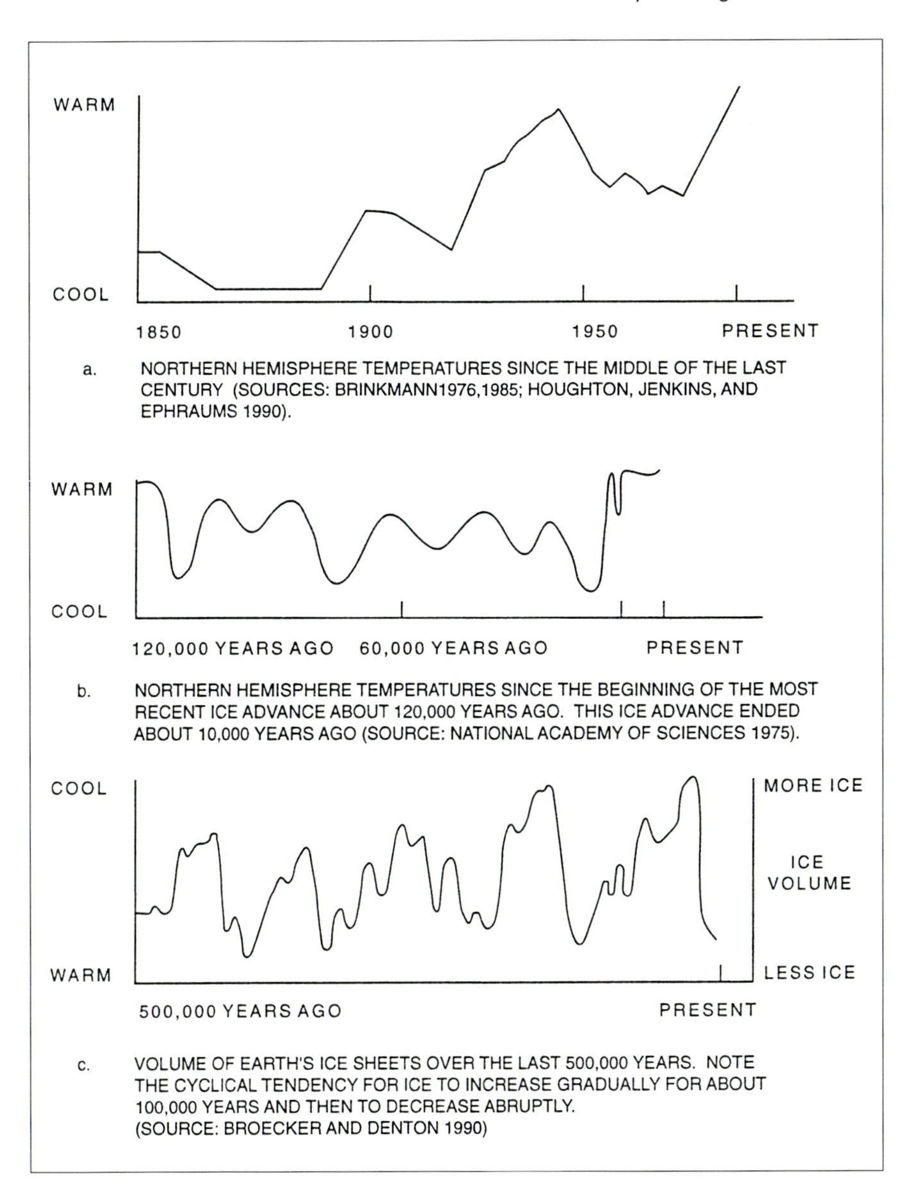

Figure 19. Climatic conditions of the past at different time scales.

cyles of streamflow and of snow accumulation and melt are the most obvious features in the Tuolumne landscape that follow cyclical patterns of change (Figure 18). With the first major snowstorm in fall or early winter, most of the Tuolumne country is buried in snow. Warming in the spring initiates melting, and bare ground emerges, rapidly at first on sites exposed to the sun and on places with only shallow accumulations, and then more slowly as the remaining snow becomes increasingly restricted to sites with shade or deep drifts, until only the bodies of more or less permanent ice and snow remain. In contrast to the area covered by snow, the depth of snow accumulation at any one locale follows a more varied climb over the winter months, with steep increases during stormy periods

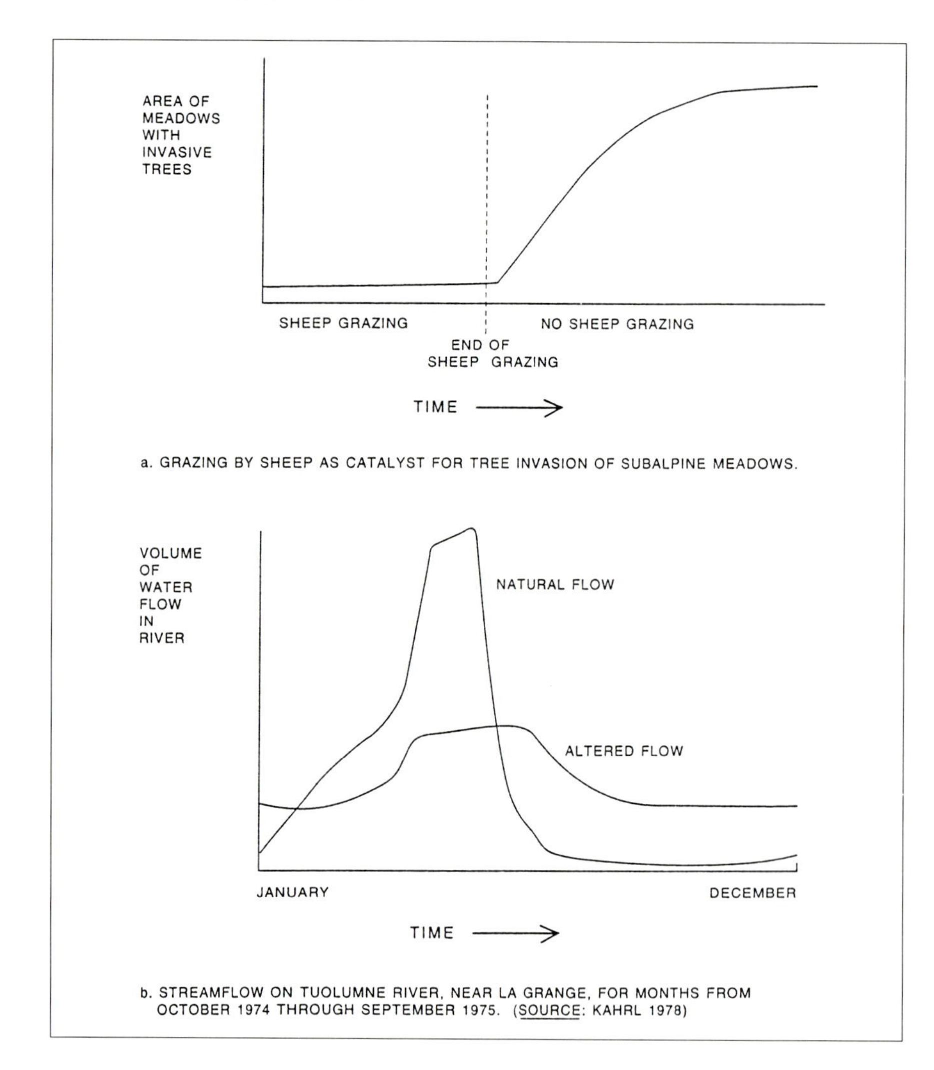

Figure 20. Directional changes in the Tuolumne landscape.

and steady or declining depths as the snow melts or compacts during stretches of quiet weather. Streamflow patterns are strongly linked to the annual cycles in weather, with maximum flows coincident with the spring snowmelt and low flows with the late fall season.

Most environments will experience such seasonal cycles in water on the landscape, but the timings will be different from place to place. In lowland California, for example, snowfall is only occasional and limited to the highest ridges, and streamflow peaks are normally in midwinter, tied to periods of heaviest rainfall. In the high mountains of southern Arizona, by contrast, streams may reach high flows during times of midsummer thunderstorms and perhaps again in spring when winter snowpacks melt. Whether in Tuolumne or elsewhere, of course, the timing and intensity of water characteristics in the landscape vary from year to year depending upon the vagaries of precipitation and temperature.

No features of natural environments change in more different ways, in fact, than weather and climate. Daily ranges of temperatures and the annual march of temperature and precipitation are strongly cyclical for the average year, although the variabilities from day to day and year to year can be large. The seasonal patterns vary greatly from place to place; California, for example, has a strong winter precipitation peak and a dry summer, whereas Wisconsin has a warm season peak and no dry season. For longer time scales, periods of distinctive weather (such as cold and wet or warm and dry) are characteristic, with either gradual or abrupt beginnings and endings to the periods, and without necessarily a repetition of the overall pattern from one period to the next (Figure 19); the beginning of the cold period of the Ice Age is an example of such a directional change. The cycles *within* the Ice Age climates suggest, however, that some climate periods may be cyclical.

Human activities are often envisioned as changing natural systems in fundamental ways, often causing directional, noncyclical change (Figure 20). Tree invasion of subalpine Sierran meadows due to sheep grazing, if true and as discussed in the vegetation chapter, is one example. The altered flow of a river, such as the Tuolumne, after it has been dammed and its water regulated or removed for human use would also reflect this sort of change. Greenhouse warming of the atmosphere, with the associated alteration of climate-dependent environmental characteristics, is still another instance. Unique, noncyclical change is often difficult to predict and the commonness of such change resulting from human activities argues that people should manipulate nature with caution, lest irreversible and undesirable consequences result (Vale 1988).

The model of a major event impelling a system in one direction rather than in another is appropriate for the life of the young boy in the family who sat beside the Tuolumne River in 1952. His life has been marked, as are all lives, by special events with long-lasting effects, those singular times when something of the future seems to hinge on a critical choice or a chance happening. Perhaps even the writing of this book reflects such an event—that 1952 visit to Tuolumne Meadows, an event solidifying a lifelong love for this place, this landscape of gently flowing water, rolling expanses of meadow sod, streamside thickets of lodgepole pine, the sloping grandeur of Lembert Dome, and the intense blue of the Western sky beyond.

AFTERWORD

Whether for people or for places, this book has been about change as well as continuity. We offer here a personal photograph as a companion piece to the one appearing in the preface.

Similarities seem at first scarce. No person in the 1950s photograph reappears in the 1980s scene. Families, like the strongest granite boulders, fragment, and the vagaries of life—and landscape—send the disjoined elements in their separate directions. Yet the routes of the journey are not random. One child from the early photograph has wandered only a short distance southward, where she makes her home in the shadows of the Sierra's rugged east face. Another lives far to the east, near a gentler coast and range, though her mind is often drawn back to the sun-basked meadows and pinnacled peaks of the Tuolumne high country. The mother rests forever in California's lowland hills amid the late winter blooms of the Indian warrior and the eternally cheerful twittering of foraging bushtits.

As with the landscape, on the broader time scale continuity generally prevails. Here the child of another generation of Vales joins her mother in a mountain reverie atop a nameless dome higher in elevation and a meadow removed from the more famous Lembert Dome framed earlier. And the photographer capturing loved ones and loved surroundings is not the 1950s father (who still enjoys an occasional drive along the Tioga Road), but the father's son. Families, like streams, may narrow or widen, but seldom does the main channel fail to reestablish its central, steady flow. And the family of all who love nature, like the snows that blanket the mountains each winter, are drawn year after year to renew their communion with Yosemite's high country, to accept changes slight or inevitable, to resist changes widespread or catastrophic—for we are the guardians of, as well as those guarded by, this priceless legacy of verdant greenery, rushing waters, and enduring granite.

LITERATURE CITED

Alpha, T. R., C. Wahrhaftig, and N. Huber. 1987. "Oblique Map Showing Maximum Extent of 20,000-Year-Old (Tioga) Glaciers, Yosemite National Park, Central Sierra Nevada, California." *Miscellaneous Investigation Series, Map I-1885*. Washington: U.S. Geological Survey.

Arno, Stephen. 1984. *Timberline*. Seattle: The Mountaineers.

Billings, W. D. 1988. "Alpine Vegetation." Pp. 391–420 in M. G. Barbour and W. D. Billings, eds., *North American Terrestrial Vegetation*. Cambridge: Cambridge University Press.

Brinkmann, W. A. R. 1976. "Surface Temperature Trend for the Northern Hemisphere—Updated." *Quaternary Research* 6:355–58.

———. 1985. "The Northern Hemisphere Temperature Curve: Representativeness and Interpretive Fallacies." *Physical Geography* 5:165–85.

Broecker, W., and G. Denton. 1990. "What Drives Glacial Cycles?" *Scientific American* 262 (1): 48–56.

Byrne, R., and E. Edlund. 1990. "Late-Quaternary Vegetation Change in the Central Sierra Nevada: Pollen, Charcoal, and Macrofossil Evidence from the Stanislaus Drainage." Paper presented at the 17th Annual Natural Areas Conference ("Natural Areas and Yosemite: Prospects for the Future"), Yosemite, Concord, and San Francisco, California.

Chorley, R., and R. Beckinsale. 1980. "G. K. Gilbert's Geomorphology." Pp. 129–42 in "The Scientific Ideas of G. K. Gilbert: An Assessment on the occasion of the centennial of the United States Geological Survey (1879–1979)," *Geological Society of America Special Paper 183*. Boulder, Colorado: Geological Society of America.

Colby, W., ed. 1960. *John Muir's Studies in the Sierra*. San Francisco: Sierra Club.

Fox, S. 1983. *The American Conservation Movement: John Muir and His Legacy*. Madison: University of Wisconsin Press.

Fritts, H., G. Lofgren, and G. Gordon. 1979. "Variations in Climate Since 1602 as Reconstructed from Tree Rings." *Quaternary Research* 12:18–46.

Gaines, D. 1977. *Birds of the Yosemite Sierra*. Oakland: California Syllabus.

Gibbens, R., and H. Heady. 1964. "The Influence of Modern Man on the Vegetation of Yosemite Valley." *Manual 36*. Berkeley: California Agricultural Experiment Station, Extension Service.

Goin, P., C. E. Raymond, and R. E. Blesse. 1992. *Stopping Time: A Rephotographic Survey of Lake Tahoe*. Albuquerque: University of New Mexico Press.

Gould, S. J. 1986. "Evolution and the Triumph of Homology, or Why History Matters." *American Scientist* 74:60–69.

Greene, L. W. 1987. *Yosemite National Park: Historic Resource Study.* Denver: National Park Service.

Gruell, G. E. 1983. "Fire and Vegetative Trends in the Northern Rockies: Interpretations from 1871–1982 Photographs." *General Technical Report INT-158.* Ogden, Utah: Intermountain Forest and Range Experiment Station, U.S. Forest Service.

Hastings, J., and R. Turner. 1965. *The Changing Mile: An Ecological Study of Vegetation Change with Time in the Lower Mile of an Arid and Semiarid Region.* Tucson: University of Arizona Press.

Houghton, J. T., G. J. Jenkins, and J. J. Ephraums. 1990. *Climate Change: The IPCC Scientific Assessment.* Cambridge: Cambridge University Press.

Humphrey, R. R. 1987. *90 Years and 536 Miles: Vegetation Changes Along the Mexican Border.* Albuquerque: University of New Mexico Press.

Kahrl, W. L., ed. 1978. *The California Water Atlas.* Sacramento: State of California.

Klikoff, L. G. 1965. "Microenvironmental Influence on Vegetational Pattern Near Timberline in the Central Sierra Nevada." *Ecological Monographs* 35:187–211.

Koerber, T. W. 1973. "Return of the Needle Miner." *Yosemite* 43 (2):3–4.

Lettenmaier, D. P., and T. Y. Gan. 1990. "Hydrologic Sensitivities of the Sacramento–San Joaquin River Basin, California, to Global Warming." *Water Resources Research* 26:69–86.

Matthes, F. E. 1930. "Geologic History of the Yosemite Valley." *Professional Paper 160.* Washington: U.S. Geological Survey.

McCaughey, W. W., and W. C. Schmidt. 1990. "Autecology of Whitebark Pine." Pp. 85–96 in "Proceedings—Symposium on Whitebark Pine Ecosystems: Ecology and Management of a High Mountain Resource." *General Technical Report INT-270.* Ogden, Utah: Intermountain Research Station, U.S. Forest Service.

Muir, J. 1894. 1961 ed. *The Mountains of California.* Garden City, New York: Natural History Library.

———. 1901. *Our National Parks.* Boston: Houghton-Mifflin.

———. 1911. 1987 ed. *My First Summer in the Sierra.* New York: Penguin Books.

———. 1912. 1962 ed. *The Yosemite.* Garden City, New York: Natural History Library.

Murie, O. 1954. *A Field Guide to Animal Tracks.* Boston: Houghton-Mifflin.

National Academy of Sciences. 1975. *Understanding Climatic Change.* Washington: National Academy of Sciences.

National Park Service. 1990. "Fire Management in Yosemite, 1970–1989." Mimeographed sheet.

Neely, W. 1958. "Tioga Peak." *Yosemite Nature Notes* 37 (12):160–63.

O'Neill, E. S. 1983. *Meadow in the Sky: A History of Yosemite's Tuolumne Meadows Region.* Fresno, California: Panorama West Books.

Parker, A. J. 1986. "Persistence of Lodgepole Pine Forests in the Central Sierra Nevada." *Ecology* 67:1560–67.

———. 1989. "Forest/Environmental Relationships in Yosemite National Park, California, USA." *Vegetatio* 82:41–54.

Phillips, F., M. Zreda, S. Smith, D. Elmore, P. Kubik, and P. Sharma. 1990. "Cosmogenic Chlorine-36 Chronology for Glacial Deposits at Bloody Canyon, Eastern Sierra Nevada." *Science* 248:1529–32.

Pyne, S. 1980. *Grove Karl Gilbert.* Austin: University of Texas Press.

Rogers, G. 1982. *Then and Now: A Photographic History of Vegetation Change in the Central Great Basin Desert.* Salt Lake City: University of Utah Press.

Rogers, G., G. Malde, and R. Turner. 1984. *Bibliography of Repeat Photography for Evaluating Landscape Change.* Salt Lake City: University of Utah Press.

Sax, J. L. 1980. *Mountains Without Handrails.* Ann Arbor: University of Michigan Press.

Sharsmith, C. 1956. "The Glacier Moraines of Lower Dana Meadows." *Yosemite Nature Notes* 35 (12):173–75.

Sherman, C. K., and M. L. Morton. 1984. "The Toad That Stays on Its Toes." *Natural History* (April): 72–78.

Smith, S., and R. S. Anderson. 1992. "Late Wisconsin Paleoecologic Record from Swamp Lake, Yosemite National Park, California." *Quaternary Research* 38:91–102.

Stebbins, R. C. 1954. *Amphibians and Reptiles of Western North America.* New York: McGraw Hill.

Sumner, L., and J. Dixon. 1953. *Birds and Mammals of the Sierra Nevada.* Berkeley, California: University of California Press.

U.S. Bureau of Land Management. 1984. *Historical Comparison Photography: Headwaters Resource Area, Butte District.* Butte, Montana: Bureau of Land Management.

Vale, T. 1987. "Vegetation Change and Park Purposes in the High Elevations of Yosemite National Park, California." *Annals of the Association of American Geographers* 77:1–18.

———. 1988. "Clearcut Logging, Vegetation Dynamics, and Human Wisdom." *Geographical Review* 78:375–86.

———. 1989. "Vegetation Management and Nature Protection." Pp. 75–86 in G. Malanson, ed., *Natural Areas Facing Climatic Change.* The Hague: Academic Publishing.

Vale, T. and G. Vale. 1983. *U.S. 40 Today: Forty Years of Landscape Change in America.* Madison: University of Wisconsin Press.

Van Wagtendonk, J. W. 1986. "The Role of Fire in the Yosemite Wilderness." Pp. 2–9 in "Proceedings of the National Wilderness Research Conference: Current Research." *General Technical Report INT-212.* Ogden, Utah: Intermountain Research Station, U.S. Forest Service.

Veblen, T., and D. Lorenz. 1991. The Colorado Front Range: A Century of Ecological Change. Salt Lake City: University of Utah Press.

Walter, H. 1973. 2nd rev. ed. *Vegetation of the Earth.* New York: Springer-Verlag.

Weaver, T., F. Forcella, and D. Dale. 1990. "Stand Development in Whitebark Pine Woodlands." Pp. 151–55 in "Proceedings—Symposium on Whitebark Pine Ecosystems: Ecology and Management of a High Mountain Resource." *General Technical Report INT-270.* Ogden, Utah: Intermountain Research Station, U.S. Forest Service.

Wood, S. 1975. Holocene Stratigraphy and Chronology of Mountain Meadows, Sierra Nevada, California. Ph.D. dissertation (Geology), California Institute of Technology.

———. 1984. "East Meadow of Aspen Valley, Western Yosemite National Park." Pp. 1–20 in S. Stine, et al., eds., *Field Trip Guidebook,* Friends of the Pleistocene, Pacific Cell.

INDEX